Social and Ecological Interactions in the Galapagos Islands

Series Editors

Stephen J. Walsh, Department of Geography
University of North Carolina at Chapel Hill
Chapel Hill, USA

Carlos F. Mena, College of Biological & Environmental Studies
Universidad San Francisco de Quito
Quito, Ecuador

The Galapagos Islands are a "living laboratory" for the study of evolution, environmental change, and conflicts between nature and society. Free of human predators for almost all its history, these islands have developed some of the most unique life forms on the planet, adapted to their harsh surroundings and living in ecological isolation. It was not until Charles Darwin's famous visit in 1835, which helped inspire the theory of evolution that this Archipelago began to receive international recognition. The Galapagos Archipelago encompasses 11 large and 200 small islands totaling approximately 8,010 sq. km. This series will focus on the entire island archipelago, and it will emphasize the study and documentation of human-environment interactions on the four inhabited islands in the Galapagos: Isabela, Santa Cruz, San Cristobal, and Floreana. Together they constitute a well-defined "natural laboratory" for the study of human-environment interactions as they vary in fundamentally important ways.

John Dunn Insua • Ximena Córdova
Jaime López-Andrade
Editors

Human Galapagos

Boundaries, Territories and Places

Springer

Editors
John Dunn Insua
Institute of Architecture and Urban Form (AFORU)
Universidad San Francisco de Quito (USFQ)
Quito, Ecuador

Ximena Córdova
College of Science and Engineering
Universidad San Francisco de Quito USFQ
Cumbayá Quito, Ecuador

Jaime López-Andrade
Institute of Architecture and Urban Form (AFORU) and Instituto de Energía y Materiales
Universidad San Francisco de Quito (USFQ)
Quito, Ecuador

ISSN 2195-1055 ISSN 2195-1063 (electronic)
Social and Ecological Interactions in the Galapagos Islands
ISBN 978-3-032-08799-7 ISBN 978-3-032-08800-0 (eBook)
https://doi.org/10.1007/978-3-032-08800-0

This Springer imprint is published by the registered company Springer Nature Switzerland AG
The registered company address is: Gewerbestrasse 11, 6330 Cham, Switzerland

Foreword 1

Declared a UNESCO World Heritage Site in 1978 and a biosphere reserve in 1985, the Galápagos Islands occupy a particular place in human imagination alongside other wonders of nature such as the Great Barrier Reef and the Serengeti savanna. Yet, apart from the ephemeral first visits to the archipelago by individuals when a pristine ecosystem without human interference could be encountered, the Galápagos Islands have been "Human Galápagos" for at least three hundred years—a landscape altered by human exploitation, settlement, and invasive species introduced by humans.

Four hundred ninety years before the publication of this book, Fryar Tomás de Berlanga, Bishop of Panama, discovered the Galápagos Islands by accidence on March 10, 1535, after his ship had drifted off course while he was sailing to Peru. In the absence of clear evidence of earlier human visitors to the archipelago, this is the beginning of human presence on the Galápagos, 300 years before Charles Darwin reached the islands on September 15, 1835. Pirates were the main visitors to the islands prior to the 1780s, introducing Black rats to the Galápagos as their most destructive invasive species.

By Darwin's times, whaling, sealing, and the massive use of giant tortoises as a source of fresh meat was leading to a precipitous and devastating decrease in the islands' characteristic species. From the records of ship logs we can reconstruct that between 1830 and 1850, 5000 Galápagos sperm whales were caught around the islands; an estimated up to 200,000 giant tortoises were removed from the Galápagos by c. 700 whaling ships between 1800 and 1870 (leading to their complete disappearance from some islands), and the Galápagos fur seals were hunted to near extinction. Over time, the survival of many of the 1300 endemic species has come under threat by more than 1600 invasive species, some of which were introduced to the Galápagos rather recently—such as the Avian vampire fly (introduced in the 1960s, in addition to rats and cats a major threat to land birds) and the hill raspberry (introduced in 1968). Colonists, specimen collectors, and bush fires ignited by humans contributed further to the loss of biodiversity.

In this context, it is disconcerting to read Herman Melville's description of the Galápagos, during the heyday of whaling, as the most desolate place on earth, as lands that could exist "in no world but a fallen one," "looking much as the world at large might after a penal conflagration," and in addition cursed by the lack of any change, "neither the change of seasons nor of sorrows." Was it not precisely his times that brought desolation, conflagration, and curse to the Galápagos archipelago?

Only in recent decades has "Human Galápagos" changed from a relationship of human exploitation and destruction of the islands to one aiming to conserve and restore them. In 1959, the Government of Ecuador declared the Galápagos archipelago a National Park (followed by the establishment of a National Park Service in 1968 and a first National Park Master Plan in 1974); in the same year, the Charles Darwin Foundation for the Galápagos Islands was established (and later other nonprofit organization dedicated to the protection of the archipelago, such as Galápagos Conservancy in 1985). UNESCO recognized the Galápagos Islands as a World Heritage Site in 1978 and as a biosphere reserve in 1985; in 1986, the 70,000 km^2 of ocean surrounding the archipelago was declared a marine reserve.

This universal commitment to the protection of the Galápagos Islands is shared by the Association of Pacific Rim Universities (APRU), the leading consortium of research universities in the Asia-Pacific, comprising currently 62 premier institutions of higher education, representing more than 2 million students and 200,000 faculty. Since its inception in 1997, APRU has been a trans-Pacific platform for collaboration that contributes effectively to the solution of major challenges of the Asia-Pacific, be it in the area of sustainability, biodiversity, multi-hazards, global health, education, or technology. Since 2016, the University of Oregon has led our key program facilitating solutions in sustainability, "Sustainable Cities and Landscapes." The present handbook on "Human Galápagos" is the collective academic outcome of the program's 6th annual conference, hosted by our Ecuadorian member Universidad San Francisco de Quito (USFQ) both on its campus in Quito and its Galápagos branch on the island of San Cristóbal, in August 2023. Building on this success, we decided to establish an inaugural Sustainable Cities and Landscapes Galápagos Summer School, held from June 22 to July 13, 2024, to offer students a unique opportunity, through three weeks of field study, to gain first-hand knowledge of the Galápagos Islands and their modern challenges.

This "little world within itself," as Charles Darwin called the Galápagos Islands, has, through human interference on the islands and human change imposed on our entire planet, been drawn into the larger transformations of a greater world from which it was secluded since the geological emergence of the archipelago. It has thus become an exemplary case of interconnectedness on Planet Earth, and the universal responsibility that humans hold for life and survival on this planet. The present volume "Human Galápagos" is a token of this responsibility and, as we hope, a building block of the path into the archipelago's future.

It is a pleasure to conclude this foreword by extending my gratitude to all those who have realized the conference and compiled the volume—Professors John Dunn, Jaime Eduardo López Andrade, and Ximena Córdova Vallejo from Universidad San Francisco de Quito, Professors Dennis Galvan (Program Chair) and Yekang Ko

(Program Director) from the APRU Sustainable Cities and Landscapes Program leadership at the University of Oregon with their support team, as well as Christina Schönleber (Chief Strategy Officer) and Benjamin Zhou (Program Manager) from the APRU Secretariat.

Thomas Schneider

Chief Executive
APRU
Hong Kong, China

Foreword 2

The Galápagos Islands stand as one of the most emblematic natural landscapes on Earth—a living laboratory of evolution and a global biological hotspot. However, the Islands are much more than their pristine biodiversity and captivating landscapes. They are a dynamic social-ecological system, shaped by the interplay of natural processes and human activity, offering a profound case study for addressing some of the most pressing challenges of our time: how to reconcile human development with biodiversity conservation, and how to envision a future where both can thrive.

This book, *Human Galapagos*, is an extraordinary outcome of the 6th APRU Sustainable Cities and Landscapes (SCL) Conference, hosted by Universidad San Francisco de Quito (USFQ) in August 2023. The conference brought together scholars, educators, students, and practitioners from across the Pacific Rim to engage in an intensive exploration of place-based approaches to tackling the social, ecological, and technical challenges facing the Galápagos. These collaborative efforts spanned disciplines, geographies, and perspectives, reflecting the core mission of the APRU SCL program: to drive innovation and transdisciplinary knowledge exchange to address local issues with global relevance.

I deeply thank USFQ Professors John Dunn, Jaime Eduardo López Andrade, and Ximena Córdova Vallejo for their leadership in successfully hosting such a valuable conference and collecting outcomes of the working groups to edit this book.

The Galápagos Islands, with their unique biodiversity and precarious balance between human activity and conservation, exemplify the tensions that arise in regions under development pressures. While the challenges may seem specific to this archipelago—such as invasive species management, freshwater scarcity, urban morphologies, and energy transitions—the lessons learned here have profound implications for coastal and island communities across the Pacific Rim and beyond. The contributions in this volume reflect a shared commitment to addressing these challenges through transdisciplinary collaboration, blending ecological science, architecture, landscape architecture, urban planning, engineering, policy analysis, and community engagement.

A Living Laboratory of Coexistence

The chapters in this volume underscore a central theme: the Galápagos is not merely a collection of islands, but a living system of coexistence. This coexistence is shaped by ecological and evolutionary forces, as well as by human ingenuity, resilience, and adaptation. From the interplay of endemic and invasive species to the energy transitions critical for climate action, the book reveals how human decisions ripple through ecological systems. It highlights that the solutions to these challenges cannot be siloed into technical fixes or policy mandates alone. Instead, they demand an integrated approach that accounts for the social, ecological, and technical dimensions of sustainability—a framework that is central to the APRU SCL's approach.

Take, for example, the energy transitions explored in this book. The Galápagos faces a unique dilemma: the urgent need to shift from fossil fuels to renewable energy, while safeguarding its fragile ecosystems and addressing the needs of its communities. This duality reflects a broader truth about sustainable development—the solutions we seek must be as complex and multifaceted as the challenges themselves. Similarly, the examination of water scarcity and wastewater management illustrates how a seemingly local issue—freshwater availability—intersects with global questions of equity, governance, and resilience in the face of climate change.

Place-Based Knowledge for Global Impact

A key strength of this book lies in its place-based focus. The Galápagos is a microcosm of the broader challenges facing humanity in an era of unprecedented environmental change. The insights presented here—whether about the urban morphologies of San Cristóbal's waterfront, the cultural imaginaries shaping conservation policies, or the design of ecological landscapes—offer models and frameworks that can be adapted and applied elsewhere. By grounding these inquiries in the specifics of place, the book bridges the gap between local knowledge and global application, demonstrating that sustainable solutions must be rooted in the local context while contributing to broader conversations on sustainability.

Transdisciplinary Collaboration: The Heart of Innovation

What makes this volume particularly remarkable is the collaborative process behind it. The APRU SCL program, with its network of 21 core member universities, is a testament to the power of global partnerships. The contributors to this book represent a diversity of disciplines, from urban design and landscape architecture to ecology and anthropology. This diversity is not just a strength—it is a necessity. The complex challenges facing the Galápagos, and indeed our world, cannot be addressed by any single discipline or perspective. It is through the integration of knowledge across fields and the co-creation of solutions with local stakeholders that true innovation emerges.

This ethos of collaboration extends beyond academia. By engaging with policymakers, practitioners, and community members, the APRU SCL program ensures that its work is not only academically rigorous but also socially relevant and actionable. The field school in the Galápagos, for instance, exemplifies how experiential learning and on-the-ground engagement can enrich research and create a new generation of sustainability leaders who are deeply attuned to the nuances of place.

A Call to Action

As you engage with the chapters of "Human Galapagos," you will encounter stories of resilience, creativity, and hope. You will see how the challenges faced by the Galápagos are, in many ways, a microcosm of the challenges facing our planet. But you will also see how collaboration, rooted in place-based understanding and guided by a vision of sustainability, can yield transformative solutions.

This book is more than a scholarly exploration of the Galápagos. It is a call to action—a reminder that the choices we make today, whether in remote islands or bustling urban centers, have profound implications for the future of our planet. It is an invitation to embrace the complexity of our world and to approach its challenges with humility, curiosity, and a commitment to equity and justice.

As the Director of the APRU Sustainable Cities and Landscapes Program, I am deeply inspired by the work presented in this volume. It reaffirms our shared mission to create resilient, equitable, and vibrant communities across the Pacific Rim and beyond. I invite you to read this book not just as a collection of research and case studies but as a vision for what is possible when we come together to imagine and build a more sustainable future.

Yekang Ko

Director
APRU Sustainable Cities and Landscapes Program
Eugene, OR, USA

Associate Professor of Landscape Architecture
University of Oregon
Eugene, OR, USA

Senior Scientist (Joint Appointment)
Pacific Northwest National Laboratory (PNNL)
Richland, WA, USA

Contents

Chapter 1
A Historical and Critical Study of Spatial Planning for Nature Conservation in the Galapagos Islands

Jaime López-Andrade, Yizhao Yang, John Dunn Insua, Andrew Boeckman, Alejandra Priera, and Diana Salazar

Abstract The history of planning in the Galapagos Islands has received limited attention in the existing Galapagos literature. While socio-ecological research has addressed issues such as introduced species, landscape transformations, and other socio-ecological interactions, few studies have specifically examined the application of different planning approaches and their impacts on human settlements and natural conservation. Given the range of management and development policy tools that have influenced the spatial distribution across the archipelago, planning studies constitute valuable resources for understanding the evolving geographical and socio-ecological contexts of the islands.

This chapter analyzes socio-ecological interactions in the Galapagos Islands from a spatial planning perspective. It examines the regulations and boundaries that have delineated protected from colonized areas. The historical review considers the regional development plans of the archipelago, the urban growth of Puerto Baquerizo Moreno, and the spatial transformation of the rural area of El Progreso on San Cristóbal Island, from the designation of Galapagos as a protected natural area up to the present. The findings are cross-referenced with a historical analysis of the

---

J. López-Andrade (✉)
Institute of Architecture and Urban Form (AFORU) and Instituto de Energía y Materiales, Universidad San Francisco de Quito (USFQ), Quito, Ecuador
e-mail: jlopez@usfq.edu.ec

Y. Yang
School of Planning, Public Policy and Management, University of Oregon, Eugene, OR, USA

J. Dunn Insua
Institute of Architecture and Urban Form (AFORU), Universidad San Francisco de Quito (USFQ), Quito, Ecuador

A. Boeckman
University of Oregon, Eugene, OR, USA

A. Priera · D. Salazar
Universidad San Francisco de Quito (USFQ), Quito, Ecuador

J. Dunn Insua et al. (eds.), *Human Galapagos*, Social and Ecological Interactions in the Galapagos Islands, https://doi.org/10.1007/978-3-032-08800-0_1

management structure of the Galapagos National Park and assessed within the context of the institutional framework dedicated to the ecological preservation of the archipelago. This research represents an initial step toward understanding the spatial interactions between human-made environments and the natural structures of the Galapagos Islands.

Keywords Galapagos spatial planning · Socio ecological planning · Galapagos territorial planning · Galapagos history

Introduction

Charles Darwin's theory of evolution addresses the relationship between organic forms and the physical geography of a territory, proposing that it is the territory itself that induces changes and mutations in organic forms, prompting their adaptation. Patrick Geddes, a biologist and planner from the late nineteenth century, expanded on Darwin's concepts by examining the relationship between humans and their territory (Welter 2002). Geddes argued that the specific characteristics of human settlements vary according to their position within a region, illustrating this idea with his "valley section" drawing, which depicts human adaptation to different territorial conditions.

Considered by many as the father of planning, Geddes also laid the foundations for systems thinking by emphasizing the interactions between the natural environment and human settlements. He argued that to genuinely understand a city, it is crucial to survey the territory comprehensively in all its natural dimensions, including systems such as geology, topography, hydrology, land cover, and animal movements, as well as the material and cultural aspects of urban form (Geddes 1915). These initial insights highlight the interactions between built and unbuilt environments as intricate spatial configurations that mutually influence each other.

Many authors have elaborated on the significance of Geddesian ideas within systems theory, including scholars focusing on Galápagos systems literature (Batty et al. 2019). Systems thinking, particularly regarding interactions between human and natural systems, has led to the recognition that these systems are coupled, reciprocally influencing each other to the point of co-evolution, whereby ecosystems and social systems develop interdependently. Currently, in the Galápagos, natural and human systems are already intricately interconnected (de Haan et al. 2019).

Liu et al. (2007), in their article on coupled human and natural systems (CHANS), categorize these relationships into three dimensions: organizational, temporal, and spatial. Organizationally, humans depend on nature for ecosystem services but simultaneously affect nature's capacity to provide these services. Temporal interactions highlight that the effects within CHANS unfold at varying rates; some impacts are immediate, whereas others require time and monitoring to become apparent.

Finally, spatially, CHANS function across multiple scales—from individual settlements to regional extents—and are interconnected. Local couplings interact with broader scales and do not necessarily align with territorial or human boundaries. Michael Batty, Luis Bettencourt and Michale Kirley in their chapter on Urban Galápagos book, expand on the CHANS theory by introducing the term CUNS (coupled urban and natural systems), emphasizing the hierarchical interactions between urban systems and natural environments (Batty et al. 2019).

The spatial division of the Galápagos Islands, which asserts that only 3% of the territory is colonized while 97% is designated as protected space, conceptually excludes the remainder from the framework of the national park. This perspective does not reflect systems thinking. However, in recent planning policies in the Galápagos, the concept of the socio-ecosystem—acknowledging the interactions between human and natural systems—has begun to emerge. This research examines the history of Galápagos planning through the lens of spatial planning, which seeks to rationally organize land uses by balancing territorial demands with environmental protection. In this article, we propose integrating the concepts of CHANS and CUNS with spatial planning, aiming to develop a practical tool for enhancing Galápagos spatial planning.

Methods and Materials: Galapagos Spatial Morphogenesis

The analysis applied in this study is an adaptation of the historical-geographical approach rooted in the British school of urban morphology, which seeks to explain the geographical structure, patterns, and character of human settlements through a systematic analysis of their constituent elements and their development over time (Kropf 2017). Traditionally, this method focuses on property division and land ownership as part of the historical morphogenesis of cities. However, in this case, the methodology was adapted not to trace the evolution of land ownership within the urban layout, but rather to examine the configuration of Galápagos' spatial division shaped by zoning practices.

To this end, we conducted a comprehensive review of Galápagos planning documents approved and published in Ecuador's Official Registry, beginning with the first environmental protection law for the archipelago in 1934 and extending to the most recent development plan issued by the Galapagos Government Council (CGREG by its Spanish acronym). The review included organic laws and regional development plans. Sources included the digital repository of the National Institute of Higher Studies, municipal archives, and, primarily, the Ecuadorian Official Registry. In total, more than thirty planning documents were analyzed to better understand the complexity of the territory and the evolution of planning processes.

In Galápagos literature, the most used spatial division is the distinction between protected and colonized areas—a framework primarily employed in conservation planning. However, as a province, the Galápagos has evolved in parallel with the development of regulatory and planning instruments at the provincial, cantonal, and parish levels. These documents have been analyzed from two complementary perspectives: first, a diagnostic perspective that includes a territorial analysis of the historical context; and second, the proposed programs for the archipelago's development or governance, which are generally considered short-term actions in most Galápagos planning documents.

All this information was analyzed to identify how spatial policies have been integrated into these documents, viewing the act of planning throughout history as a means of shaping the current spatial mosaic of the Galápagos. Policies affecting spatial division were examined through a cross-scale approach, integrating diverse elements into a single, nested framework to better understand cross-scale relationships. The accumulation of spatial interactions at the city level both influences and is influenced by regional-scale dynamics. Likewise, socio-ecological effects at the regional or global level undoubtedly impact local scales. To more effectively address the complexities of spatial planning in the Galápagos, this chapter employs the concepts of territory, spatial components, areas, and zones.

A territory is defined as a geographic space shaped by social, ecological, and cultural relationships. A spatial component is a broad term encompassing both defined and undefined geographic spaces, which include various types of spatial divisions. Areas are spaces within a specified perimeter, delineated either physically or conceptually. Finally, zones are segments of space characterized by specific, generally homogeneous attributes aligned with land regulation and administration. A territory can be subdivided into diverse spatial components, some formally designated as areas or zones, and others that, despite lacking official classification, can still be geographically identified and possess distinct characteristics.

In this chapter, we examine the cross-scale interactions among the city of Puerto Baquerizo Moreno, the rural area of El Progreso, the island of San Cristóbal, and the broader Galápagos Ecoregion, not as isolated case studies, but as components of an interconnected, cross-scale analysis. At the regional scale, regional development plans were considered; at the city scale, the current Urban Development Plan was reviewed. At the property or building-block scale, specific classifications such as built footprint were identified, though without applying the full set of variables typically used in urban morphological studies.

Empirical research involved the reconstruction of cadastral maps using available graphic and photographic data retrieved from historical documents and aerial photographs provided by the Geographic Military Institute of Ecuador (IGM). This data were integrated through Geographic Information Systems (GIS) to consolidate historical cadastral records with a range of geographic, urban, and rural information layers obtained from CGREG, IGM, and Charles Darwin Foundation CDF. San Cristóbal Island was selected as a case study for examining cross-scale spatial configurations, due to the historical evolution of interactions between its highland and coastal zones.

The first successful colonization of the Galápagos Islands occurred in the highlands of San Cristóbal with Manuel J. Cobos's establishment of Hacienda El Progreso (1879). Throughout history, various urban–rural connections have linked the highlands and the coast: the construction of the San Cristóbal Railroad by Cobos to transport goods to and from the hacienda; the presence of the U.S. Army, which built a pipeline to bring water from the highlands and distribute it to other islands; and the abandonment of rural land in favor of tourism development. Each of these processes has exerted pressure on natural systems through changes in land cover for agriculture, the introduction and spread of invasive species, and the exploitation of natural resources.

On the other hand, the natural elements that intersect with the built environment across the region are immediately visible, primarily in the form of streams that flow from the highlands to the coast, passing through the city of Puerto Baquerizo Moreno. In other words, CHANS and CUNS on San Cristóbal Island have taken on a visible form that has evolved over time in response to changes in rural and urban settlements. In contrast, on Santa Cruz and Isabela Islands, these same cross-scale interactions are more difficult to perceive, as they occur through less visible or more transient systems—such as the role of underground crevices in Santa Cruz's hydrological system or the migratory paths of tortoises intersected by agricultural activity in rural areas. With a population of approximately 7000, San Cristóbal represents an intermediate scale between Santa Cruz (22,000 inhabitants) and Isabela (2500 inhabitants).

Fragmentation and Construction of Galapagos Territorial Framework

Spatial zoning in the archipelago began when it was first declared a national park. Between 1934 and 1937, the earliest protection laws designated several islands as natural parks while recognizing private property rights for settlers who could provide proof of tenancy; however, these laws had limited lasting impact. In 1959, the declaration of the Galápagos as a National Park further reinforced the territorial division between protected and colonized areas, designating the entire archipelago as protected, except for urban and rural zones. This process was consolidated over the following 20 years, culminating in 1979.

Before this zoning definition was finalized, the archipelago was declared a province in 1973, prompting its division into cantons and parishes, second- and third-order administrative divisions, respectively. The Galápagos now comprises one province, three cantons, five rural parishes, and three urban parishes (Fig. 1.1). Each canton contains several rural parishes and at least one urban parish that functions as its cantonal capital.

The first management plan for the national park was developed in 1974 with the support of UNESCO, and the province's initial development plan was established in 1975. Together, these two documents created an administrative framework that corresponded to both the zoning of the park and the zoning of the inhabited areas. Collectively, the plans divided the territory into 13 distinct spatial components, where the protected areas were subclassified according to conservation-oriented

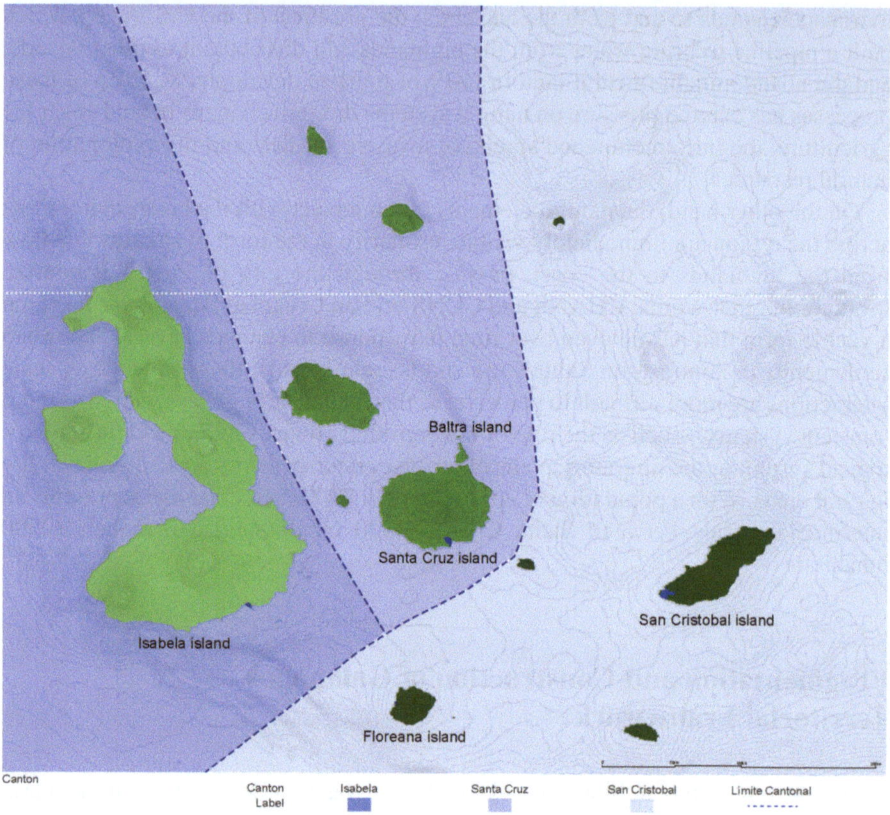

Fig. 1.1 Galapagos province division

zoning, and the colonized areas were categorized only as urban or agricultural, without acknowledging other types of land use within them (Table 1.1).

By contrast, the most recent National Park Management Plan, published in 2014, along with the currently active development plan, classifies the territory into a total of 54 spatial components. These areas are now subdivided into land-use zones that integrate both human and conservation functions. Human-occupied zones are further categorized based on the level of consolidation or the designated vocation of the territory (Table 1.2).

Over the years, spatial planning in the Galápagos has grown increasingly complex due to the growing recognition of human presence in the archipelago. In recent decades, the Galápagos has been conceptualized as a complex socio-ecosystem that necessitates the integration of social development and conservation (González et al. 2008; Quiroga 2019; Tapia et al. 2009; Watkins 2008). Since 1959, the region has produced numerous planning documents, including management, development, and regulatory plans at the park, provincial, cantonal, and parish levels. However, some of these documents have remained in effect for less than five years, reflecting a continuous yet fragmented planning process marked by a lack of consistent long-term objectives.

Table 1.1 Galapagos spatial components in 1975

Spatial entities 1975				
Area		Land use zones		
Galapagos archipelago	Protected area	Primitive scientific zone	This zone includes areas that remain unaltered and are essentially free of introduced species. They require the highest level of protection, and no human use that introduces or causes any damage to the ecosystem will be permitted	
		Primitive zone	Includes nearly all uninhabited areas of the islands, as well as shallow marine waters	
		Extensive use zone	These zones will include areas with outstanding features of high visitor interest but which cannot support heavy visitor loads	
		Intensive use zone	These are carefully selected areas, limited in total extent, of high visitor interest, where the largest number of visitors will be directed to observe the island's ecological ensembles	
		Special use zone	This zone includes substantially altered areas located near the settled areas outside the park. Certain activities are permitted, such as firewood collection, salt extraction, and the gathering of introduced fruits. Roads and trails will also be allowed within this zone	
	Colonized area	Buffer and protection zone	Land	The first 100 meters of the colonized area adjacent to the park boundaries
			Sea	The first two nautical miles from the coast, within the inland waters of the archipelago
		Rural	Agricultural and livestock	
		Urban	Urban settlements	

Source: Galapagos Management Plan 1974, Galapagos development plan 1975

Following the declaration of the Galápagos as a national park and the establishment of its boundaries in 1979, the Galápagos National Institute (INGALA) was created in 1980 to oversee provincial planning. In 1986, the entire internal sea of the archipelago was designated a marine reserve under exclusive state control, based on Ecuador's 1971 declaration of territorial waters, which defined the internal waters

Table 1.2 Galapagos spatial components 2020

Management + Development plan spatial entities						
Area subdivision			Land/water use zones			
Galapagos archipelago	Protected area	Galapagos national park GNP	Zone of absolute protection			
			Zone for restauration and conservation of ecosystems			
			Zone of impact reduction	Network of sites for public use and ecoturism	Visting sites for restricted ecotourism	
					Visiting sites for intensive ecotourism	Natural and intensive ecotourism
						Close and intensive ecotourism
						Managed intensive ecotourism
					Visting sites for recreational ecotourism	
					Cultural and educational	
				Especial public use	Sites for infrastructure and logistics	
					Sites for especial service use	
			Transition zone			
		Galapagos marine reserve GMR	Multiple use			
			Limited use	Zone for comparison and protection		
				Zone for conservation and non-extractive use		
				Zone for conservation, extractive and non-extractive use		
				Temporary special management zones		
			Port zones			

	Colonized area	Rural	Future urban expansion	Promotion of production	Presence of natural and human threats
				Restauration and mitigation	
			Extractive use	Non extractive use of resources	Presence of hydrological recharge zones, nesting areas of animals with heritage value, problems of property division
				Extractive use of resources	
			Productive use	Agriculture	
				Cattle	
				Forestry	
			Ecological protection	Ecology protection	
				Forestry preservation	
		Urban	Consolidated urban land	Consolidation	Presence of natural and human threats
				Comprenhensive improvement	
			Unconsolidated urban land	Consolidation	
				Development	
				Comprenhensive improvement	
			Protected urban land	Conservation and recovery	
			Urban land for especial uses	Industrial use	
				Artisanal use	

Source: Galapagos Management plan 2014, Galapagos Management Plan 2020 of the archipelago by drawing straight lines connecting the outermost points of the islands to form a polygon.

The 1988 development plan outlined three priorities: (1) consolidating the marine reserve, (2) developing special legislation for the Galápagos, and (3) promoting eco-social development. Two of these goals were achieved within the following decade. In 1991, the Galápagos Tourism Management Plan recognized the spontaneous economic shift from fishing and agriculture toward tourism services, leading to increased urban development. This plan identified 140 visitor sites across the islands, expanding on the 43 sites listed in the 1974 park management plan, and integrated cultural and recreational activities into the park's original zoning.

Today, the archipelago hosts 176 visitor sites. In 1996, the park's management plan incorporated new land-use zones, aligning them with the diverse visitor sites and permitted activities. This approach was reinforced in the 1998 plan, which consolidated the marine reserve and formally recognized port activities as part of it.

The 2003 development plan introduced a refined spatial classification of the Galápagos, organizing it into terrestrial, marine, and human areas. These areas were further divided into zones, acknowledging their interconnections and emphasizing that both conservation and development rely on the integrated environmental management of these spatial components.

The Concept of Socio-ecosystem in Galapagos Zoning

In 2005, the updated management plan for the Galápagos National Park recognized the archipelago as an ecoregion—a large, functional spatial unit with sufficient geographic scope to support ecological and evolutionary processes operating within its boundaries over time. It also defined Galápagos as a socio-ecosystem, characterized as a complex ecological system interconnected with social systems and subdivided into distinct subsystems (Instituto nacional Galápagos & Fundación Charles Darwin, 2005).

The plan introduced a new zoning framework based on UNESCO's Man and the Biosphere Program, of which Galápagos has been a member since 1984. Utilizing this framework to define the relationships between human populations and other components of the park, the plan emphasized the three core functions of a biosphere reserve: (1) conservation, (2) development, and (3) logistics (Bioret et al., 1998).

Additionally, it adopted spatial planning principles commonly applied in biosphere reserves, which typically include a core conservation area, a surrounding buffer zone, and a transition zone, with human settlements situated outside the transition zone (Bioret et al. 1998) (Fig. 1.2). In Galápagos, however, most human activities are centrally located, with protected areas encircling these anthropogenic zones. The 2005 management plan recognized this distinctive configuration and proposed a model consisting of core zones—pristine ecosystems; buffer zones—areas immediately adjacent; and transition zones—spaces designed for cooperation with local populations. All these elements function within an integrated territorial framework (Fig. 1.3).

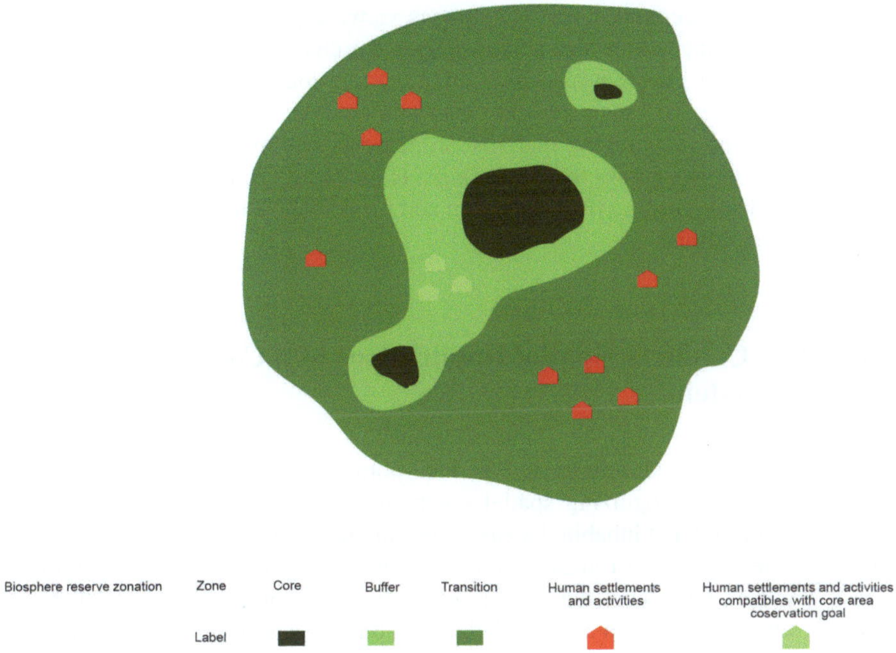

Fig. 1.2 UNESCO's Typical territorial division for biosphere reserves

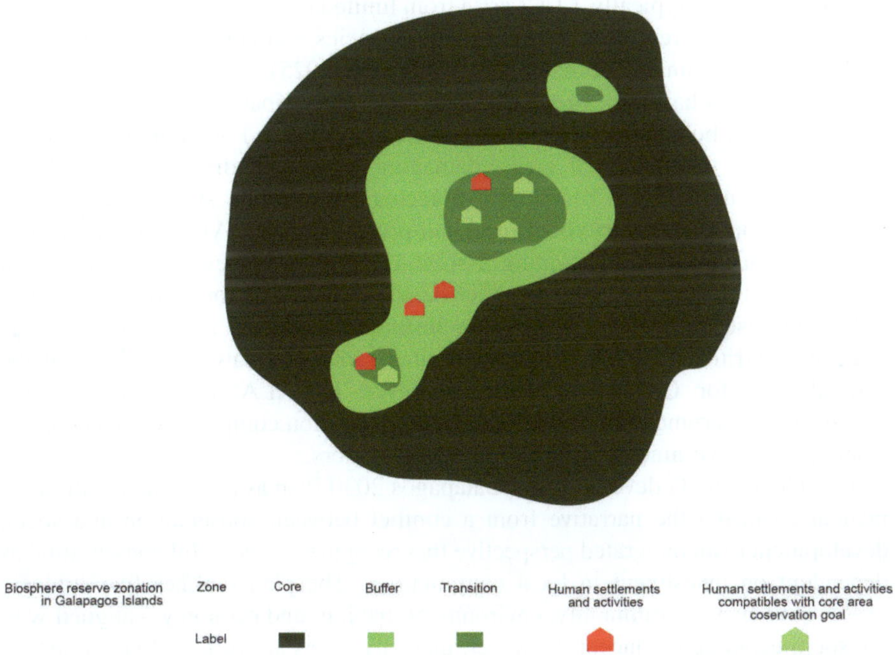

Fig. 1.3 Galapagos re-interpretation of UNESCO's territorial division for biosphere reserves

While this conceptual model encourages viewing protected and non-protected areas as interconnected, it also acknowledges the challenges posed by integrating multiple levels of governance, which often lead to conflicts over jurisdiction and resource use. The 2014 management plan reinforced the concept of Galápagos as a socio-ecosystem, recognizing that spatial elements within human areas influence ecological processes across the archipelago. However, its zoning strategy remained confined to distinguishing between urban and rural areas, without achieving deeper systemic integration.

Land Use Conflicts, Spatial Implications of a Complex Socio-ecosystem

The 2015 Provincial Development Plan incorporates the concept of Galápagos as a socio-ecosystem, recognizing spatial components that transcend the boundaries between protected and inhabited areas. These include water recharge zones in rural areas, species migration routes, and streams extending from the highlands to the coast. The plan introduces the concept of Territorial Synthesis Units (UST, by its Spanish acronym), defined as natural elements that span multiple land-use zones or fall within a single administrative category but influence the characteristics of other zones and involve multiple institutions. In planning literature, these overlapping spatial elements and administrative conflicts are commonly referred to as Land Use Conflicts (LUCs). Typically, LUCs arise from limited land availability and its varied suitability for different uses, exposing discrepancies and contradictions that pose challenges to sustainable development (Wang et al. 2025).

Such conflicts have also been identified in recent Galápagos research, highlighting that spatial boundaries are not always clearly defined. For example, giant tortoises migrate through rural areas, damaging crops and dispersing introduced species from agricultural zones into protected areas, while streams function as biotic corridors that traverse both rural and protected zones (Valdivia et al. 2014). The 2015 plan advocates adopting a socio-ecosystem framework for provincial development; however, it does not offer detailed guidelines for second- and third-order administrative levels, which have historically evolved independently of an integrated territorial vision. Additionally, that same year saw the revision of the Special Law for Galápagos, which dissolved INGALA and established the Galapagos Government Council CGREG, an institution composed of stakeholders from various governmental and administrative sectors.

In 2020, CGREG developed the Galápagos 2030 Plan as part of its institutional mandate, shifting the narrative from a conflict between conservation and social development to an integrated perspective that recognizes successful conservation as dependent on investment in local communities. The plan outlines five strategic axes—governance, community, environment, habitat, and economy—aligned with the socio-ecosystem concept. However, these five axes are further divided into 18 policies, 42 strategies, and more than 100 projects and actions, complicating their practical implementation.

Despite these complexities, the plan has stimulated research in areas such as decarbonization, food sovereignty and security, and sustainable urban planning. This is exemplified by the Symposium on Inclusive Decarbonization and the Conference on Sustainable Cities and Landscapes, both held in Galápagos in 2023. Lastly, in 2022, a new marine reserve was established to connect the Galápagos Marine Reserve with the Cocos Reserve. This expansion, which has broader regional implications, is not analyzed in this document as it exceeds the territorial scale of the Galápagos.

In addition to its organization around strategic axes, a notable strength of the Galápagos 2030 Plan is its emphasis on land-use planning at the canton and parish levels, where it establishes specific guidelines and standards. The plan subdivides urban and agricultural areas into zones based on land use and suitability. It recognizes the zoning framework defined in the 2014 park management plan and integrates it with the zoning of urban and agricultural areas. Together, these two plans divide the territory into 54 spatial components, several of which feature overlapping uses and spatial elements that cross administrative boundaries, thereby generating areas of Land Use Conflict (LUCs) (Table 1.2). These conflicts have arisen in Galápagos due to shifting land use and land cover patterns and call for dynamic, integrated management strategies to ensure their effective resolution.

Land Use Changes and Conflicts in San Cristobal Island

A clear example of land-use changes processes and the resulting conflicts can be observed in the analysis of urban and rural areas on San Cristóbal Island. According to the 2014 Management Plan, the island covers a total of 55,709 hectares, of which 733.6 hectares are designated as urban. Adjacent to the town of Puerto Baquerizo Moreno lies a rural area that connects urban and agricultural zones and has been designated as an urban expansion zone. The urbanization of this area could potentially divide the protected area into two segments, creating a barrier that fragments the transitional ecosystem across the southwestern part of the island unless development proceeds within an ecological urbanism framework.

Additionally, a significant portion of the 733 hectares classified as urban corresponds to the naval base and remains undeveloped. Since this land is under military jurisdiction, it represents a potential site for future infrastructure development. Therefore, it is essential to begin evaluating the possible environmental and spatial consequences of such development.

To analyze the spatial structure of San Cristobal Island, a distinction is drawn between the concepts of 'urban area' and 'consolidated urban zone.' In Galapagos, urban areas are defined as territories undergoing significant environmental changes due to the construction of buildings, infrastructure, roads, introduction of new species, pollution, and other human activities, which are largely concentrated in these areas (DPNG 2014). Such urban areas are defined by precise geographic boundaries established through ministerial agreements. Nonetheless, urban areas in Galapagos may also include vacant land, undeveloped zones, protected zones, and rural areas intended for future growth.

According to Ecuadorian Land Use regulations, the consolidated zone is that which, within the limits of the urban area, shows signs of a development process, possesses a minimum infrastructure endowment and can be the object of future consolidation. As opposed to undeveloped zones, which despite being within the urban limits, lack infrastructure and their territory should be transformed to be considered habitable, protected urban zones are those that due to its ecological or cultural characteristics present features that deserve to be preserve through specific policies. To analyze the growth of the consolidated zone of San Cristobal Island, we worked with the planimetry of five different years for which we were able to obtain graphic or photographic information: 1952, 1972, 1982, 2005 and 2022. The cadastral plans were reconstructed from the available information, mapping the number of existing constructions in the different years to establish the growth of the built footprint of Puerto Baquerizo Moreno. Measuring growth through changes in the urban footprint provides a clearer understanding of urban evolution than merely examining boundary shifts. In Galapagos, urban boundaries are periodically redefined and often incorporate areas designated for potential growth; therefore, calculations of density and land-use changes typically lack precision.

On the other hand, population data from censuses in the year's closest to those in which they were drawn were used to establish a relationship of density per urbanized area and built footprint per inhabitant. Between 1959 and 2022, the built footprint of Puerto Baquerizo Moreno went from 10,034.38 m2 built on the ground floor for a population of 782 inhabitants to a built footprint of 344,135.00 m2 for a population of 7290 inhabitants. In other words, there were originally 12.83 m2 of ground floor-built area for each inhabitant and in 2022 this ratio increased to a total of 47.21 m2 of built footprint for each inhabitant of Puerto Baquerizo Moreno. Similarly, the consolidated zone in 1959 was 8.41 ha, which gives a population density of 93 inhabitants/ha, the footprint built on the ground floor in 1959 is equivalent to 11.93% of the consolidated zone. On the other hand, in 2022 this proportion varied, the consolidated urban area was 212.80 ha that housed a population of 7290 inhabitants, resulting in a density of 34 inhabitants/ha with a built footprint that represents 16.17% of the consolidated zone.

This relationship shows that over the years, land use efficiency in Puerto Baquerizo Moreno has decreased, achieving lower densities with greater urbanization and construction. Likewise, the number of buildings constructed per inhabitant went from a ratio of 0.12 to 0.37 buildings constructed per inhabitant, and buildings went from an average of 99 m2 per building to an average of 125 m2 in 2022 (Table 1.3). This process of growth of the consolidated area had overlapped with natural elements that are integral part of the natural hydrological systems, the city of Puerto Baquerizo Moreno is crossed by 3 streams that connect the internal waters or the island with the sea, creating a LUC that had become a risk management due to flood hazards in the city during rain seasons (Figs. 1.4 and 1.5).

Table 1.3 Urban land use efficiency in San Cristobal Island

Urban area	Description	1959	1972	1982	2007	2022
	Built footprint[a] (m2)	10,034.38	21,232.32	42,483.70	1,86,037.00	3,44,135.00
	Growth of built footprint (%)	0	111.60%	100.09%	337.90%	84.98%
	Number of buildings (u)	101.00	186.00	275.00	1365.00	2743.00
	Average built footprint (m2)	99.35	114.15	154.40	136.29	125.45
	Consolidated urban area[a] (ha)	8.41	14.92	28.99	116.52	212.80
		IGM 1959	PDOT 1973–1975	IGM 1982	IGM 2007	GOOGLE EARTH 2024
	Urban area (ha)	10.30	16.81	88.94	153.91	422.17
		IGM 1959	PLAN 1975	PDOT 2015–2020–YEAR 1981	PDOT 2015–2020–YEAR 2006	PDOT SAN CRISTOBAL 2020–2024
	Inhabitants in Puerto Baquerizo Moreno	782.00	1311.00	1665.00	6672.00	7290.00
		Census 1962 (cited 2015 census)	Census 1974	1982 census	2010 census	Census 2022
	Population growth (%)	–	67.65%	27.00%	300.72%	9.26%
	Built urban density (m2/inhab)	12.83	16.20	25.52	27.88	47.21
	Population density in consolidated zone (inhab/ha)	92.98	87.87	57.43	57.26	34.26
	Population density urban area (inhab/ha)	75.92	77.99	18.72	43.35	17.27
	Built-up area in ground floor (%)	11.93%	14.23%	14.65%	15.97%	16.17%
	Number of properties					2921.00
	Plots with ecological service area of streams (u)					206.00
						7.05%
	Number of buildings in ecological service area of streams (u)					125.00
						4.56%
	Total area of protection of streams in urban area (ha)					23.82
						11.19%

[a]The built footprint and consolidated zone area does not take into account the airport area

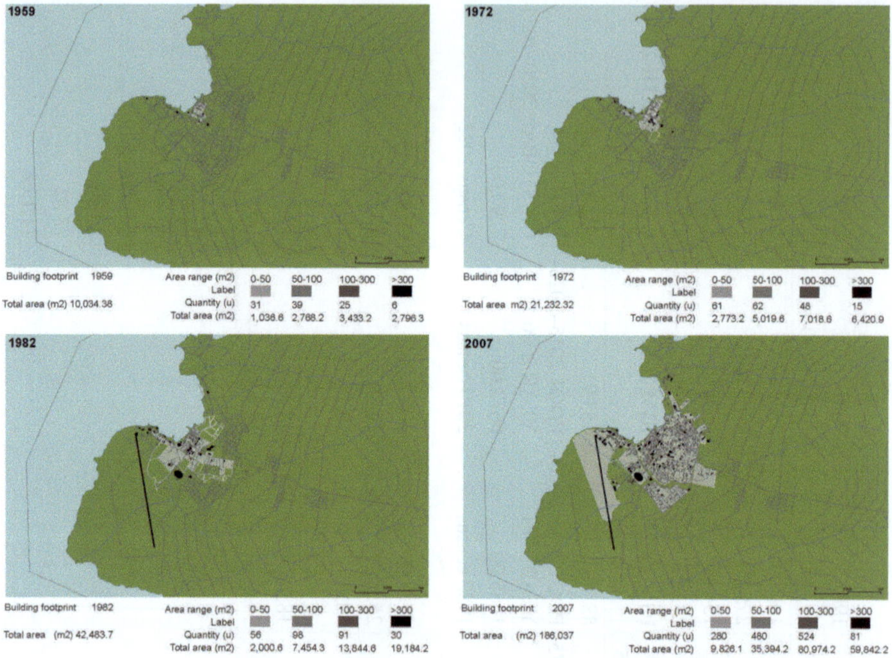

Fig. 1.4 Puerto Baquerizo Moreno, footprint growth between 1959 and 2007

Fig. 1.5 Puerto Baquerizo Moreno's 2022 Footprint and Land Use Conflicts in the Urban Area of San Cristóbal

The 2009 urban development regulatory plan, prepared by the Association of Municipalities of Ecuador AME and published in the Official Registry on August 31, 2009, establishes zones for the protection of watercourses corresponding to mandatory setbacks of 10 meters from the edge of the watercourse towards the interior of the city and sectors of influence for the protection of the watercourse edge of 50 meters measured from the edge of the watercourse, which must be taken into account from the moment the body of water crosses the urban boundary along the entire length until it empties into the sea. However, there are currently a total of 206 properties, and 125 buildings built on these protected zones; this LUC represents more than 12% of the current urban consolidated zone of San Cristobal urban area (Table 1.3).

On the other hand, because agriculture has primarily developed within the humid ecosystems of the larger islands—and since this type of ecosystem is found exclusively on these islands—there are Zones of Ecological Importance located within agricultural areas. These zones hold conservation and tourism value, as well as areas of natural vegetation. This situation gives rise to a different type of Land Use Conflict (LUC), where important habitats exist within private properties that shelter vulnerable species and are of ecological, landscape, or tourism interest.

Examples include private lands intersected by species migration routes, farms with petrel nesting sites, wetlands, and areas containing native or endemic vegetation (Benitez-Capistros et al. 2019). In San Cristóbal, such species have been identified in at least 9.09% of the rural area. Some of these sites, although privately owned, integrate conservation efforts with productive activities such as agroforestry and ecotourism.

The transformation of rural areas is difficult to measure in terms of urbanization levels, primarily because many rural zones lack infrastructure coverage. However, comparisons can be made based on the number and size of properties within the same territory, illustrating variations in land tenure. The earliest data on rural properties come from documents produced by the Central University of Ecuador as part of an urban development plan for the Galápagos Archipelago, carried out in November 1972, as part of the preparatory actions for the 1975 development plan (Junta Nacional de Planificacion y Coordinacion Económica 1975). This document recorded 253 farms distributed across 7602.91 hectares. Currently, there are 1678 farms within an area of 8235.50 hectares. Moreover, in 1972, only 138 plots were smaller than 10 hectares, whereas now there are 1354 plots of this size, primarily concentrated in the southwestern region of San Cristóbal's rural area (Table 1.4). In this case, the increased density of properties in the same area reflects not land-use efficiency but a potential Land Use Conflict (LUC).

The fragmentation of rural properties diminishes productive capacity; when rural lots are subdivided into dimensions comparable to urban plots, it signals the onset of rural urbanization. This process significantly alters land use, reduces productivity, and affects land cover, ultimately impacting both rural and protected areas. This study demonstrates that the highest levels of property subdivision occur in the southwestern region, particularly near the rural settlement of El Progreso parish (Figs. 1.6 and 1.7). Additionally, the rural area of San Cristóbal encompasses a key

Table 1.4 Rural fragmentation between 1972 and 2022

Rural properties	Description	1972	2022
	Rural area (ha)	7602.91	8235.00
		PDOT 1975	DGNP Management Plan 2014
	Number of properties	253.00	1678.00
	Number rural property (0–10 ha)	138.00	1354.00
	Number rural property (10–50 ha)	80.00	256.00
	Number of rural properties (50–100 ha)	17.00	39.00
	Number rural property (100-300 ha)	16.00	24.00
	Number rural land (+300 ha)	2.00	4.00
	Average rural land (ha)	30.00	9.70
	Inhabitants	703.00	1010.00
		Census 1974	Census 2022

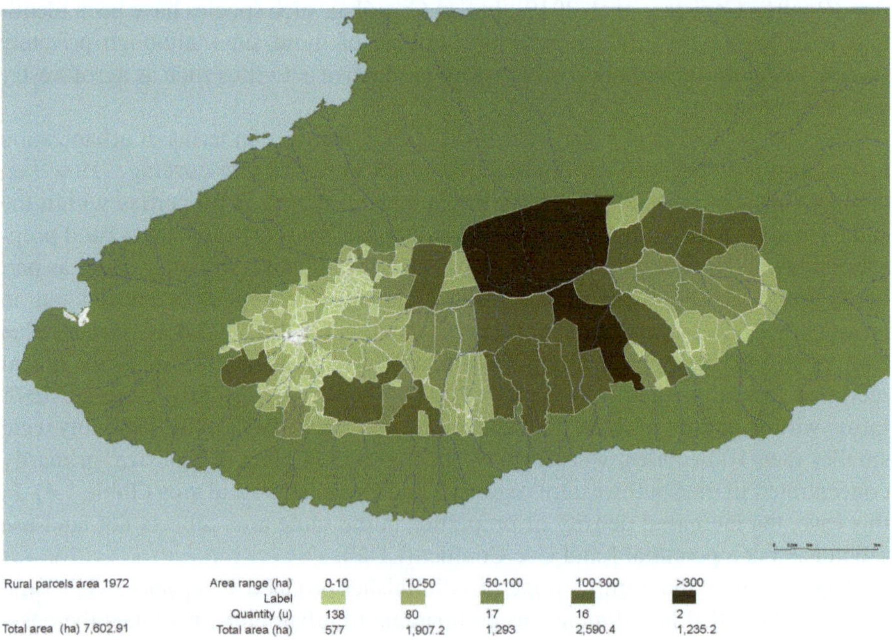

Fig. 1.6 Rural land use tenure of San Cristóbal Island in 1972

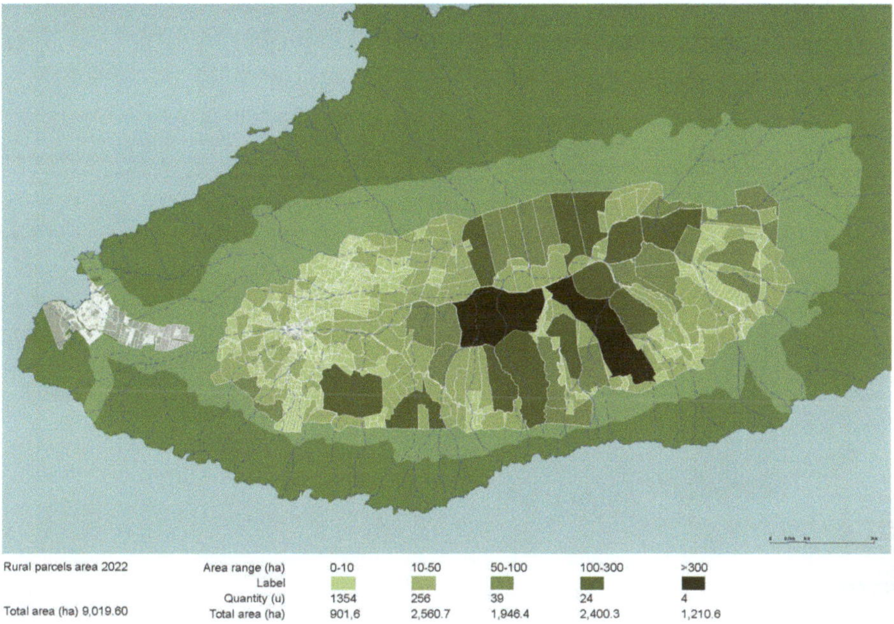

Fig. 1.7 Rural land use tenure of San Cristóbal Island in 2022

water recharge zone for the island's aquifer and springs, which supply both rural and urban populations. Many rural properties contain natural features within their boundaries, such as streams or petrel nesting habitats that cross various spatial classifications (see Fig. 1.8).

Land use and land cover transformations are dynamic and not permanently fixed; therefore, planning efforts must also be dynamic, accounting for ongoing interactions between human settlements and protected areas. Most properties larger than 50 hectares are in the eastern portion of the rural area. These properties intersect with spatial components such as streams, canyons, springs, water bodies, and aquifer recharge zones. These natural features extend beyond rural boundaries into protected areas, forming an ecologically indivisible territory that is nevertheless administratively partitioned.

Call for Innovation in Galapagos Spatial Planning, Coupled Land Use Zones and Collaborative Planning

The current spatial division acknowledges conflicts among rural, urban, and protected natural areas, as well as their respective zones. These overlapping and interacting spatial components are termed in Galapagos planning instruments as Territorial Synthesis Units (UST in Plan Galapagos 2020) or overlapping layers (in

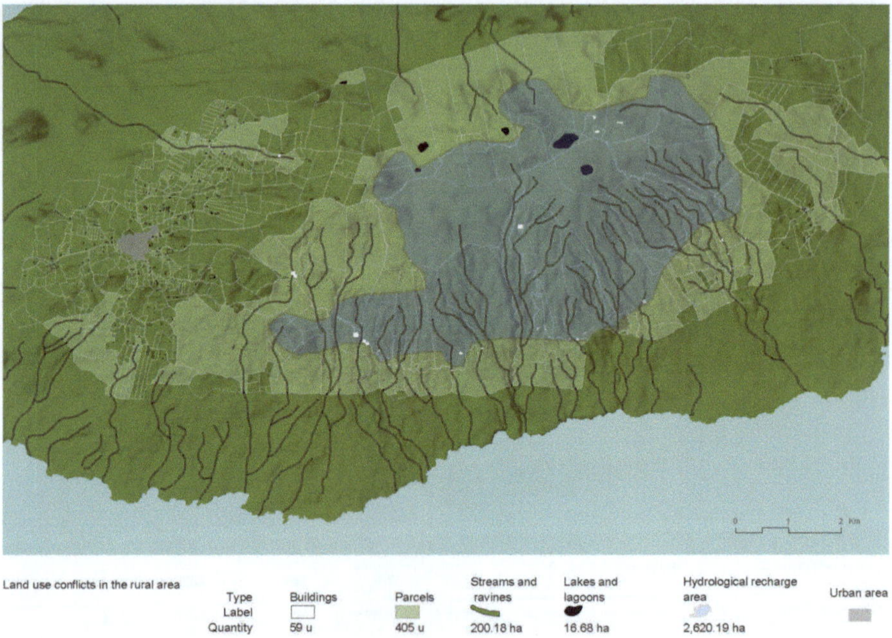

Fig. 1.8 Land use conflicts in rural areas in San Cristobal

Plan Galapagos 2030), highlighting the predominant land uses alongside layers of conflict. Although land use zoning aims to facilitate management by dividing territories into clear units, conflicts arise when these divisions do not align with natural or social systems—especially when the interactions between human and natural systems are inadequately considered (Rotich 2012). Land use conflicts are inherently multi-scale, occurring at various levels from broad geographic regions to individual human settlements.

Michael Batty, Louis Bettencourt and Michael Kirley, in the book Urban Galapagos, introduced the concept of Coupled Urban and Natural Systems (CUNS), elaborating upon the broader Coupled Human and Natural Systems (CHANS) framework. The CHANS concept analyzes interactions between human systems and their impacts on natural systems, including climate and geological processes, through ecosystem service alterations (Chen and Liu 2014). In Batty, Bettencourt and Kirley's approach, CUNS features three key properties: (1) Interactions between urban and natural systems are typically unbalanced, with human systems often controlling or heavily influencing natural systems. (2) Innovations in human systems are necessary to foster more synergistic exchanges that enhance natural system health and ensure long-term system dynamics. (3) Urban systems should pursue interactions with natural systems that are either absent or that enhance resource flows and biodiversity (Batty et al. 2019).

Meanwhile, CHANS emphasizes spatial coupling, recognizing that human-nature interactions extend beyond political or ecological boundaries and occur at multiple scales (Liu et al. 2007). Building on these concepts, one possible approach to addressing Galapagos Land Use Conflicts (LUCs) could involve creating integrative spatial components, proposed here as 'Coupled Land Use Zones' (CLUZ). This concept would address LUCs by promoting synergistic interactions between overlapping or intersecting land uses, fostering mutualistic relationships within spatial components, and thus benefiting the entire system.

For CLUZ to be effective, it will require interdisciplinary research and long-term monitoring from diverse perspectives, including scientific, social, and design disciplines such as ecology, sociology, and planning. This approach necessitates innovative management practices, technological solutions, and infrastructure interventions, and depends on collaborative spatial planning among various levels of government responsible for different land uses. The proposed Coupled Land Use Zones (CLUZ) would specifically emerge in contexts where multiple land-use classifications overlap spatially or where particular spatial components cross multiple zones—for example, recharge areas within rural zones or streams traversing urban areas—thereby linking the various zones defined within the Galápagos spatial planning system (Fig. 1.9).

Examples of collaborative management exist in rural areas of Galapagos, with private properties cooperating alongside the Galapagos National Park Directorate to maintain ecological services. Nevertheless, the scale and number of recognized LUCs surpass existing successful practices, especially in urban-rural-ecological

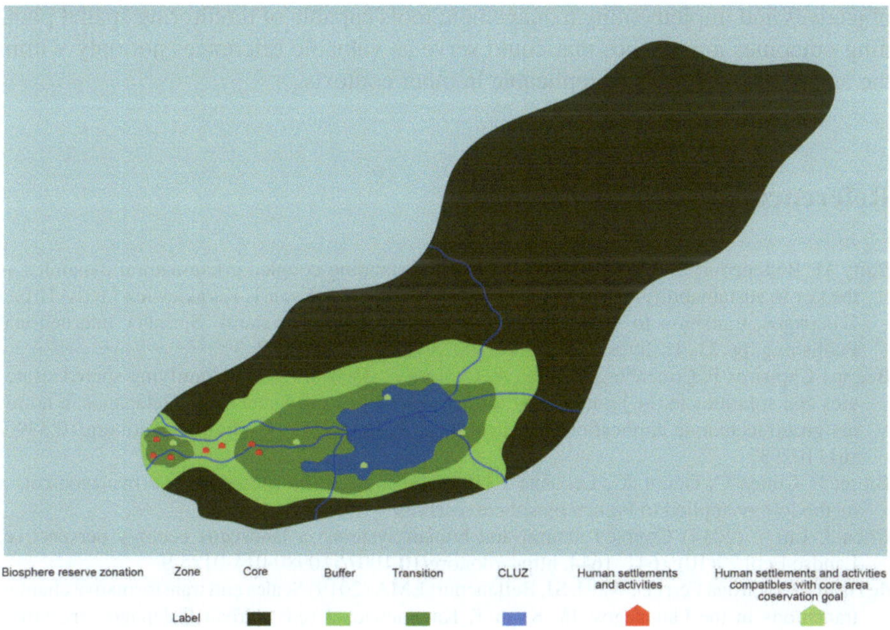

Fig. 1.9 Diagrammatic integration of CLUZ within the conceptual spatial distribution diagram of the Galapagos, using San Cristobal Island as an example

interfaces. Additionally, the analyzed documents lack clear definitions and specified tools necessary to systematically address the demand for co-management, highlighting a significant gap in current land management frameworks.

Conclusions

In Galapagos, certain socio-ecological interactions, such as species introductions, have been consistently monitored. Programs aimed at eradicating or mitigating the impacts of invasive species have existed for over 60 years. However, spatial processes resulting from planning have lacked similar monitoring, leading to the development of human settlements that feel disconnected from the conservation objectives of the archipelago. Consequently, there is a need for enhanced monitoring to ensure planning processes achieve intended socio-ecological outcomes.

In this context, a planning practice that integrates politics, the city, and the region, considering human beings as an integral part of their environment, provides an opportunity to rethink planning in Galapagos toward an authentically socio-ecological spatial definition. Likewise, Galapagos could benefit from frameworks that emphasizes bio-cultural dynamics and fosters social interactions that maintain or enhance biodiversity while supporting human well-being.

Thus, Galapagos should leverage the established concept of socio-ecosystem and position itself as a model for territorial research. This would entail setting long-term objectives and implementing management tools capable of monitoring spatial planning outcomes approaches that could serve as valuable references not only within the archipelago but also be replicable in other contexts.

References

Batty M, Bettencourt LMA, Kirley M (2019) Understanding coupled urban-natural dynamics as the key to sustainability: the example of the Galapagos. In: Kvan T, Karakiewicz J (eds) Urban Galapagos, transition to sustainability in complex adaptive systems. Springer International Publishing, pp 23–41. https://doi.org/10.1007/978-3-319-99534-2_3

Benitez-Capistros F, Couenberg P, Nieto A, Cabrera F, Blake S (2019) Identifying shared strategies and solutions to the human-giant tortoise interactions in Santa Cruz, Galapagos: a nominal group technique application. Sustainability (Switzerland) 11(10). https://doi.org/10.3390/su11102937

Bioret F, Cibien C, Génot JC, Lecomte J (1998) A guide to biosphere reserve management: a methodology applied to French biosphere reserves

Chen J, Liu Y (2014) Coupled natural and human systems: a landscape ecology perspective. Landsc Ecol 29(10):1641–1644. https://doi.org/10.1007/s10980-014-0125-9

de Haan FJ, Quiroga Ferri D, Walsh SJ, Bettencourt LMA (2019) Scales and transformative change: transitions in the Galapagos. In: Kvan T, Karakiewicz J (eds) Urban Galapagos, transition to sustainability in complex adaptive systems. Springer International Publishing, pp 43–58. https://doi.org/10.1007/978-3-319-99534-2_4

DPNG (2014) Plan de Manejo de las Áreas Protegidas de Galápagos para el BUEN VIVIR

Geddes P (1915) Cities in evolution. Williams & Norgate. https://ia800302.us.archive.org/12/items/citiesinevolutio00gedduoft/citiesinevolutio00gedduoft.pdf

González JA, Montes C, Rodríguez J, Tapia W (2008) Rethinking the Galápagos Islands as a complex social-ecological system: implications for conservation and management. Ecol Soc 13(2):13

Instituto nacional Galápagos, & Fundación Charles Darwin (2005) Plan de Manejo del Parque Nacional Galápagos

Junta Nacional de Planificacion y Coordinacion Económica (1975) DESARROLLO DE GALAPAGOS 1973–1975 (p. 368). http://repositorio.iaen.edu.ec/handle/24000/1584

Kropf K (2017) The handbook of urban morphology. Wiley

Liu J, Dietz T, Carpenter SR, Folke C, Alberti M, Redman CL, Schneider SH, Ostrom E, Pell AN, Lubchenco J, Taylor WW, Ouyang Z, Deadman P, Kratz T, Provencher W (2007) Coupled human and natural systems. Ambio 36(8):639–649

Quiroga D (2019) Socioecological systems and the management of the natural resources in the Galapagos. In: Kvan T, Karakiewicz J (eds) Urban Galapagos: transition to sustainability in complex adaptive systems. Springer International Publishing, pp 85–93. https://doi.org/10.1007/978-3-319-99534-2_6

Rotich D (2012) Concept of zoning management in protected areas 2(10). www.iiste.org

Tapia W, Ospina P, Quriga D, Reck G, González JA, Montes C, Cruz E, Milstead B, Wolff M, Watkins G, Carrión C, Piu M, Oviedo M (2009) Toward a shared vision of Galapagos: the archipelago as a socioecological system. Galapagos Report 2007–2008

Valdivia G, Wolford W, Lu F (2014) Border crossings: new geographies of protection and production in the Galápagos Islands. Ann Assoc Am Geogr 104(3):686–701. https://doi.org/10.1080/00045608.2014.892390

Wang D, Wang M, Zheng W, Song Y, Huang X (2025) A multi-level spatial assessment framework for identifying land use conflict zones. Land Use Policy 148. https://doi.org/10.1016/j.landusepol.2024.107382

Watkins G (2008) A paradigm shift in Galapagos research. Galapagos Research, (Formerly Noticias de Galápagos) 65(65):30–36

Welter V (2002) Biopolis: Patrick Geddes and the city of life. MIT Press

Chapter 2
Waterfront "Membranes" and Port Infrastructure Development in the Galápagos Islands: The Case of Puerto Baquerizo Moreno

John Dunn Insua, Jaime López-Andrade, Antonella Mercedes Segovia Amores, and Danna Leane Espinosa Alvarez

Abstract Waterfronts represent a physical boundary between land and water masses, but also function as "membranes," fluid zones of transition that regulate both ecological composition and urban development. In the Galápagos Islands, waterfronts simultaneously connect and isolate residents, to and from both the mainland and other islands in the archipelago. These relations are ever evolving, as port infrastructure, its usage and governance, and overarching contexts have shifted over time. In this chapter, we apply the "waterfront membrane" framework to a historical case study of Puerto Baquerizo Moreno—the capital of Galápagos Province and the main port on San Cristóbal Island. Our theoretical framework highlights the sophisticated organization and vulnerabilities of urban waterfronts, and we offer critical future areas of investigation relevant to the Galápagos Islands and other coastal states with similar fragile ecosystems.

Keywords Waterfronts · Port city infrastructure · Sustainability · Galápagos Islands

---

J. Dunn Insua (✉)
Institute of Architecture and Urban Form (AFORU), Universidad San Francisco de Quito (USFQ), Quito, Ecuador
e-mail: jdunn@usfq.edu.ec

J. López-Andrade
Institute of Architecture and Urban Form (AFORU) and Instituto de Energía y Materiales, Universidad San Francisco de Quito (USFQ), Quito, Ecuador

A. M. S. Amores · D. L. E. Alvarez
Universidad San Francisco de Quito (USFQ), Quito, Ecuador

J. Dunn Insua et al. (eds.), *Human Galapagos*, Social and Ecological Interactions in the Galapagos Islands, https://doi.org/10.1007/978-3-032-08800-0_2

Insular Coastlines: A Geographical Membrane

Territories form the foundation of regional planning, with environmental and geographical dynamics playing key roles in urban development. Beyond physical features, biological components are also essential in defining territorial landscapes. This chapter explores how these elements intersect in the Galápagos Islands, an archipelago where the balance between human development and environmental conservation presents significant challenges.

Insular ecosystems are intricately shaped by external environmental forces and geographical features. Flora and fauna typically reach islands through sea currents and prevailing winds. In the case of the Galápagos Islands, most of its living creatures were carried by the Cromwell, Humboldt, and Panama currents (Moore 2021). Prior to human introduction of non-native and invasive species, geographic isolation played a critical role in determining which species became part of an island's ecosystem. However, the shoreline continues to act as the ultimate ecological filter.

Shorelines function as natural boundaries where land meets ocean, creating a transitional zone that influences both ecological composition and urban development. On an insular scale, this boundary operates like a permeable *waterfront membrane*, regulating interactions between terrestrial and marine ecosystems. At the continental scale, the waterfront membrane is less significant in shaping biological environments, as the sheer mass of land facilitates interactions among and between inner regions and, consequently, multiple ecosystems. However, at an insular scale, the waterfront membrane becomes critically important. The smaller the island, the fewer the opportunities to develop distinct ecosystems within its territory. Thus, most native island biodiversity must successfully cross waters and overcome any challenges posed by an island's coastline.

Waterfronts' geographical characteristics are highly relevant when studying the territorial dynamics of the Galápagos Islands, whether from an environmental or urban perspective. Geological history is arguably the most influential factor, and oceanic islands can be classified as continental, barrier, coral, tectonic, or volcanic. In case of the Galápagos, the islands were formed by a volcanic hotspot in the Nazca plate, known as the Galápagos Hotspot (Moore 2021). This hotspot is continually reshaping the archipelago (Carrión-Mero et al. 2024). Younger islands are defined by rocky coastlines of solidified magma and, over time, erosion forms more permeable shores—mainly beaches of rocks or sand (Gouldie 2001). The convergence of a waterfront's characteristics and traits of species who journey by sea to shore defines which will survive and settle in this new volcanic environment.

Permeability: From Biochemistry to Geography

In biochemistry, a membrane is understood as a thin, semi-permeable barrier, boundary, or lining. More specifically, a membrane is "a thin physical interface through which certain species are passed or filtered according to their physical/

chemical properties" (Jang 2023). The interface porosity defines the type of species that can permeate.

Coastlines and waterfronts are a physical boundary between two states of the matter: liquid and solid. Geographically, this change requires species to possess specific traits if attempting to cross. For example, a cliff face with violently crashing waves may present significant obstacles to a reptile with claws, like iguanas or tortoises. Quadrupeds with hooves may find it nearly impossible to pass. Pinnipeds, like seals or sea lions—while better evolved to pass between land and sea—would face similar challenges and traverse better across sandy or rocky beaches. However, that same cliff face would be far less daunting to a flying seabird and even offer advantages as a safe nesting location away from predators.

Similar examples can be drawn across varied geographical features, but intuitively, depending on evolutionary histories and physical characteristics, species will be better suited to traverse different types of coastlines. Therefore, coastlines serve as geographical "membranes," and the "porosity" (i.e., geographical features of the waterfront) filters which species are established within each island's interior ecosystems.

Situating in the Galápagos Islands

Here, we illustrate our argument that an island's ecological composition is linked to the permeability of its waterfront membrane through a comparative analysis of two distinct islands in the Galápagos archipelago: Darwin Island and San Cristóbal Island.

In addition to varied locations, size, and age, the two islands exhibit vastly different coastlines. Darwin, one of the smallest islands in the archipelago, is surrounded by rocky cliffs and ravines (Fig. 2.1). Accordingly, the island is only inhabited by bird species. Contrastingly, San Cristóbal's coastline is more permeable, featuring rocky and sandy beaches (Fig. 2.1) that reptiles and mammals can traverse to reach the interior of the island.

Of course, it is also worth considering that Darwin Island is one of the most isolated islands in the archipelago, which could contribute to its species composition. However, similar biodiversity is present on Kicker Rock—another island with an "impermeable" waterfront membrane—which is near San Cristóbal and much closer to the mainland. Both Darwin Island and Kicker Rock are home to mainly birds, rather than the mammals and reptiles present throughout the archipelago. We argue that these ecological similarities between the islands are primarily defined, not by their proximity to other landmasses or water currents, but by their similar coastlines. This case study further signals the relevance of coastlines in shaping the islands' interior ecosystems.

Fig. 2.1 Comparison of waterfront permeability on Darwin Island and San Cristóbal Island. Blue stars and yellow outlines reflect permeable surfaces (e.g., sand, beaches), and red outlines represent less permeable features (e.g., rock, cliffs). (Graphic: John Dunn)

Living by the Membrane

Geography influences island ecosystems and, together, these dynamics shape the territories where humans settle.

Humans have historically sought hospitable waterfronts, characterized by sandy beaches and sheltered bays where ships can anchor safely. It is in these places where we tend to establish our ports; the human infrastructure that helps us to pass people and goods through the coastal membrane. These locations often also meet additional

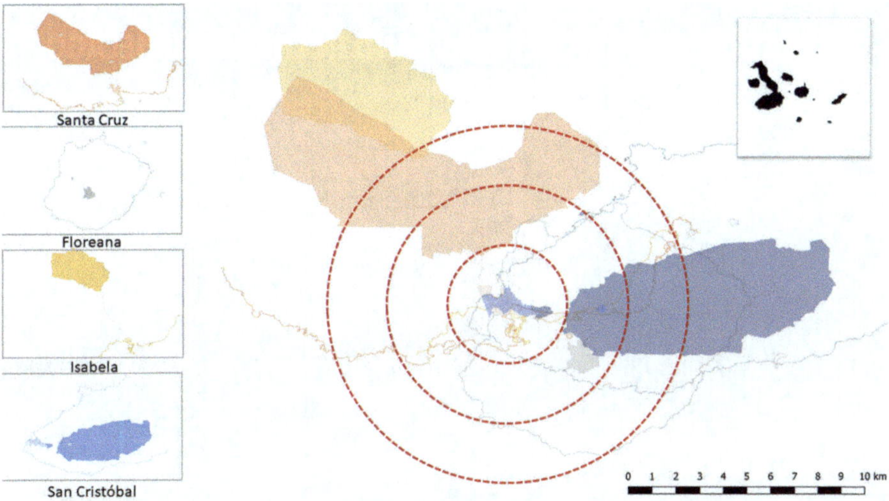

Fig. 2.2 Distance analysis from all ports in the Galápagos to respective farmlands. When the maps of inhabitated islands are overlapped on their ports' centroid, it's esay to see how similar their distances between ports and farmlands are (Graphic: John Dunn)

criteria, including access to freshwater and proximity to arable land for agriculture and livestock. When analyzing human settlements in the Galápagos, these patterns are evident. Settlements are generally positioned within manageable distances to farmland (Fig. 2.2), with slopes between these points tending to be gentle rather than steep.

In all of the inhabited islands (i.e., San Cristóbal, Santa Cruz, Floreana, and Isabela), farmlands were established at approximately 160 meters above sea level (Fig. 2.3). Most are also within 5 kilometers from their respective port settlements, with the farmlands of Isabela being the only exception of this rule at 15 kilometers away from Puerto Villamil (Fig. 2.2). The exception in this case, may be explained by the great extension of the island.

Of course, the permeability of the chosen waterfront must allow easy management of people and cargo from the ships to the land and vice versa. That implies a series of coastal rearrangements within these settlements over time. The following sections of this chapter apply the "waterfront membrane" framework to a historical case study of Puerto Baquerizo Moreno, the capital of Galápagos Province and the main port on San Cristóbal.

Guayaquil: Sourcing Goods Beyond the Membrane

Since the Galápagos Islands were annexed as an Ecuadorian territory in 1832, the archipelago has relied on the large port city of Guayaquil as its primary.

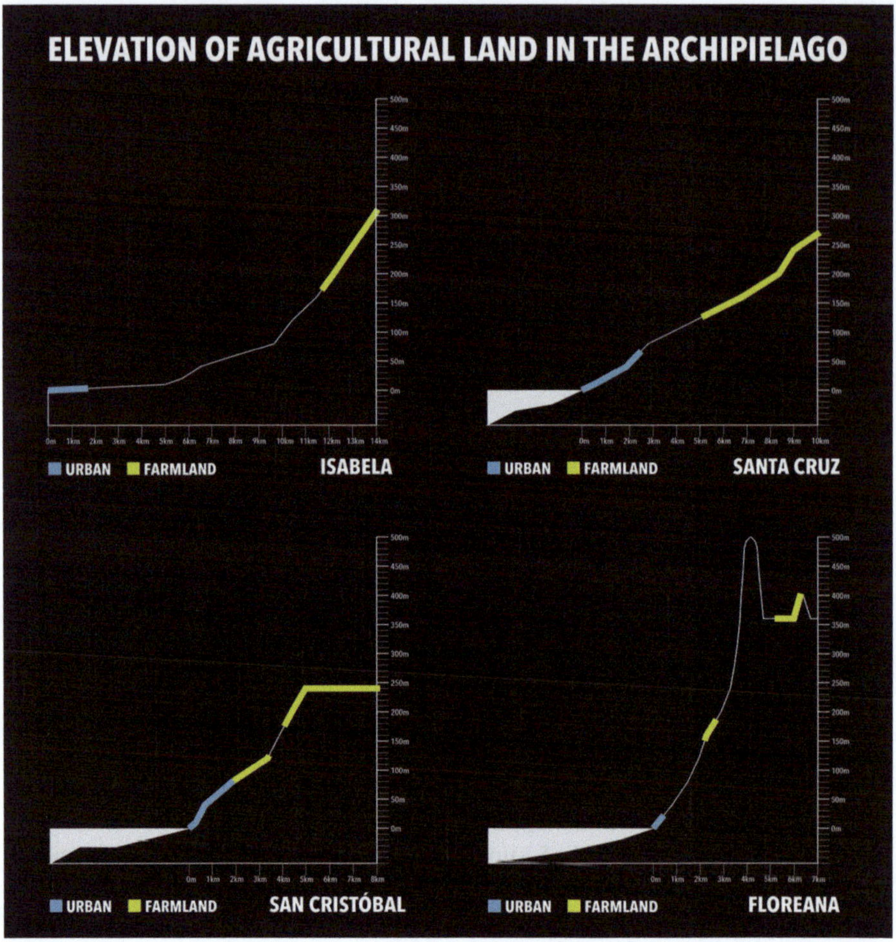

Fig. 2.3 Altitude of urban settlements (blue) near the coastline and farmlands (green). Most of the inahbited islands have their farmlands at 160 meters, due to weather conditions

connection to the Ecuadorian mainland (Latorre 2011). As analyzed in "Galápagos 2030: Galápagos Islands Strategic Plan," Guayaquil remains the most important linkage (Fig. 2.4). The city contains two ports where the ships set sail to the archipelago, full of merchandise and supplies (Redacción Plan V 2020). These ships arrive most frequently in Puerto Baquerizo Moreno and Puerto Ayora, followed by Puerto Villamil and Puerto Velasco Ibarra (Plan Galápagos 2030 2012).

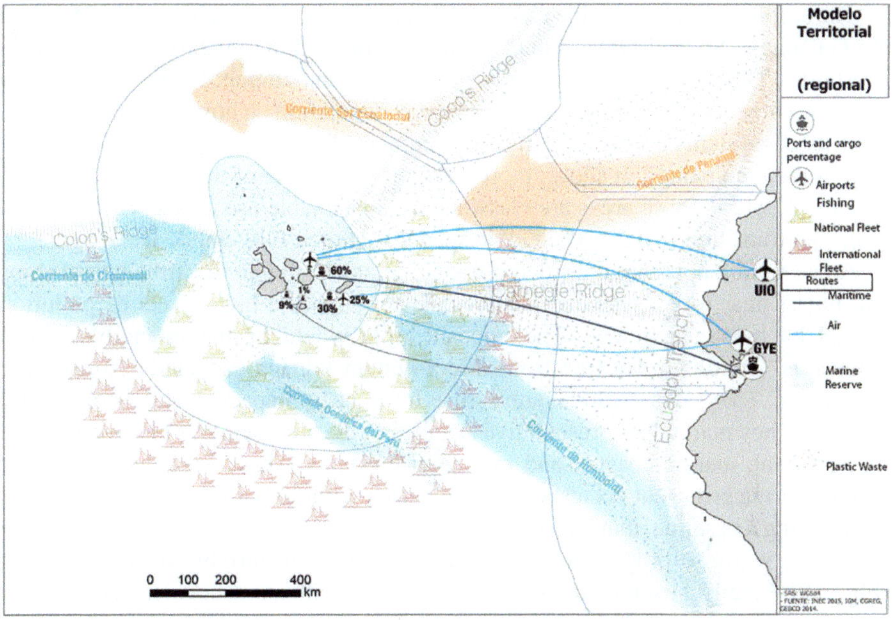

Fig. 2.4 Territorial dynamics around the archipelago. Connection routes between the islands and the mainland. Also, places used by fishign fleets. (Graphic: Plan Galápagos 2030)

The Inner Maritime Boundaries: Imaginary Membranes

Since the establishment of Galapagos' Marine Reserve, in 1998. there was a basic zoning, based on its use and purpose. Nevertheless, these boundaries were not mapped publicly for a long time. Their specific measurable boundaries may be found in some academic papers, but they haven't been shared massively for the population's knowledge. The only well-known definition of boundaries out to sea was the one between the Marine Reserve and international waters. In 2017, the Ecuadorian Navy captured and retained the "*Fu Yuan Yu Leng 999*"; a Chinese fishing ship that trespassed into the Galapagos Marine Reserve for fishing with no permit (Sarango and Paucar 2017). Later, that ship was confiscated by the Ecuadorian Government and became part of the Ecuadorian Navy Fleet, v6re-baptized as the "Hualcopo". Now it is used to protect the marine fauna of the Galapagos and keep away ships interested in commit illegal fishing in protected Ecuadorian waters (Buque chino pasó de llevar pesca ilegal a cazar depredadores en Galápagos 2024).

The events occurred with the *Fu Yuan Leng 999* evidenced how the limits of Galapagos Marine Reserve also work as a membrane. Its nature is more political than geographical. Fishing activities are forbidden in the marine reserve; and therefore, it keeps the international fishing fleets away. Otherwise, some other boats may run with the same fate as the *Fu Yuan Leng 999*.

The ambiguity of the marine reserve's inner organization lasted until 2025, when the Galapagos Special Government Council published an annex to its Galapagos 2030 plan. This document concentrates mostly in land zoning. However, it applies the same categories inside the Marine Reserve, depending on how protected or damaged its ecosystems are. Although this document does not show a comprehensive map of the archipelago, it shows more detailed the zoning of the inhabited islands, including Baltra. These areas are classified as: sustainable use, conservation, intangible and transition (Consejo de Gobierno del Régimen Especial de Galápagos).

Areas nearby ports are considered transition zones, due to the affluence of ships and boats, near to ecosystems that prevail, although the high levels of pollution caused by human activities. The opposite pole would be the intangible zones; defines as "pristine areas free from direct human impact, especially from exotic organisms. They also include areas that have been impacted and ecologically restored. They maintain representative samples of the archipelago's biodiversity and its terrestrial, marine-coastal, and marine ecosystems. These areas have levels of ecological integrity that correspond to scenarios without human intervention, thus maintaining a baseline, that is, they are reference sites. Conservation zones are those which are worth to protect, regardless the presences of introduced fauna or flora (Fig. 2.5). Finally, sustainable use areas are linked with some coexistence between nature preservation and food production (Consejo de Gobierno del Régimen Especial de Galápagos).

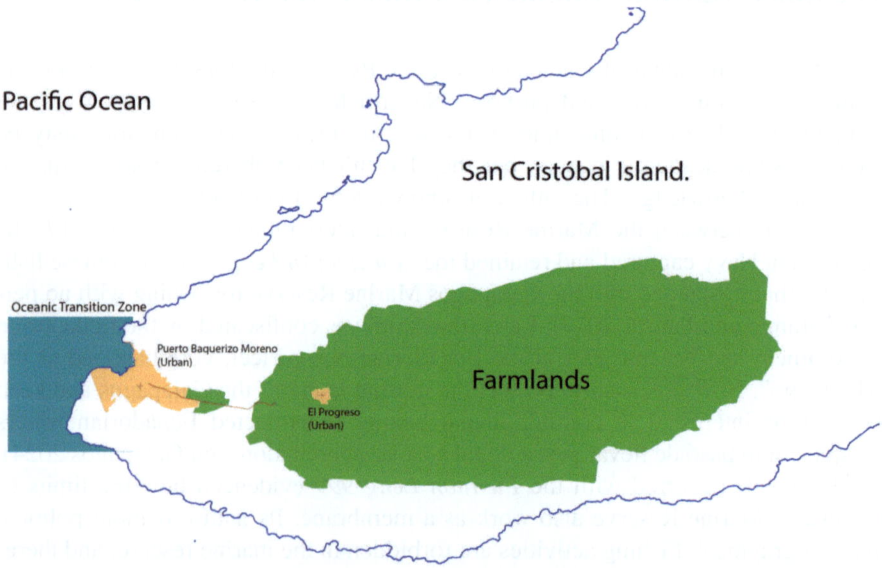

Fig. 2.5 Partial map of San Cristobal island, showing urban and rural zones. It also shows its Oceanic Transition Zone, near Puerto Baquerizo Moreno (image: John Dunn Insua)

However, beyond the accuracy of the marine zoning specified in the annex of the Galapagos 2030 plan, these new inner boundaries do not have the strength to work as political membranes and split human activities overseas; as the Marine Reserve does.

Waterfront Infrastructure in Puerto Baquerizo Moreno: Developing the Membrane

The coastal membrane also affects the way in which urban settlements relate with the ocean and with what is beyond it. As previously said, bays that may protect anchored ships with easy access to farmable lands is the preference of coastal villages that later turned into small urban settlements. That is also the starting point for the port cities in Galapagos, including the island in San Cristobal.

Besides all the published information found during this research, two interviews were conducted for this work: one with César Hernandez, the first Governor of the Galápagos Province in 1979, and another with Whitman Cox, a local enthusiast of Galápagos history.

While it is true that the Galápagos Islands were discovered in 1534, human occupation was intermittent until the end of the nineteenth century. The first attempt to establish a permanent settlement occurred on Floreana Island. This island served for a long time as the political headquarters of the Ecuadorian government, until Manuel J. Cobos—an Ecuadorian entrepreneur, cascarilla oil extractor, and sugar cane producer—recommended that the maritime governor relocate his office to San Cristóbal. While Floreana had only intermittent access to freshwater, San Cristóbal had the great advantage of a permanent freshwater source.

The first settlement on San Cristóbal was in "El Progreso," Cobos's place of residence and the camp for his workers, located in the inlands. His need to supervise shipments of products to the Ecuadorian mainland pushed him to build a wooden dock and a tiny waterfront settlement in Wreck Bay, first known as "Puerto Chico" (Fig. 2.6). Later, the village was renamed to "Puerto Baquerizo Moreno" in honor of Alfredo Baquerizo Moreno, the first Ecuadorian president to visit the islands (Idrovo 2013).

As a membrane, the coast of Wreck Bay is predominantly impermeable, except for four points with sandy beaches: those points will be later known as: Playa de los Marinos, the former largest beach in the area; Playa Dorada; Playa Mann and the beach near Punta Carola, a little further. Cobos pier was built in Playa de los Marinos, splitting the beach in half. Interestingly, long after Cobos was killed by his on workers in a riot, the dock built by him was later used by the U.S. Navy to transport freshwater from San Cristobal to their Base Beta—nicknamed "The Rock"—on Baltra Island in the archipelago. The American base was built to defend the Panama Canal from a possible Japanese attack during World War II (Idrovo 2013). Following the war, in the 1970s, the dock was demolished.

Fig. 2.6 Photograph of Wreck Bay and Cobos's wooden dock. (Photo: American Expedition of 1936)

During the last years of the Ecuadorian military dictatorship, during the 1970s, local residents attempted to build a new port in Puerto Baquerizo Moreno. These attempts failed, largely due to a lack of construction materials and poor workmanship. Both interviewees mentioned that the debris from the collapsed docks with some other dragged by the rains from the island remain underneath the sands of Playa de los Marinos, the location of Cobos's original dock, explaining why that beach is not recommended for swimmers.

As the need to improve Puerto Baquerizo Moreno's connectivity with the other islands and the mainland, four docks have been constructed over the last 50 years. This infrastructure increased the permeability of the coastal membrane.

One one of hi piers is solely for use by the Ecuadorian Navy. A second—the longest one—was first designed for cargo management and fishing, and then later remodeled for tourist and local transportation around the archipelago. The third, known as "Aquarius Dock," was used first for tourism and transport and is now exclusively used by cruise ships. Lastly, the "Fishermen's Pier" was built during the first decade of his century. Ironically, this pier is not exclusively for fishermen. It is often used by ships that require some mechanical maintenance.

Reviewing aerial photographs from Ecuador's Instituto Geográfico Militar (Geographic Military Institute), Puerto Baquerizo Moreno's growth and evolving morphology is evident (Fig. 2.7). The archive has images from 1959, 1982, 1993, and 2003. Importantly, the photographs show how the growth of the town is linked to the development of its port infrastructure.

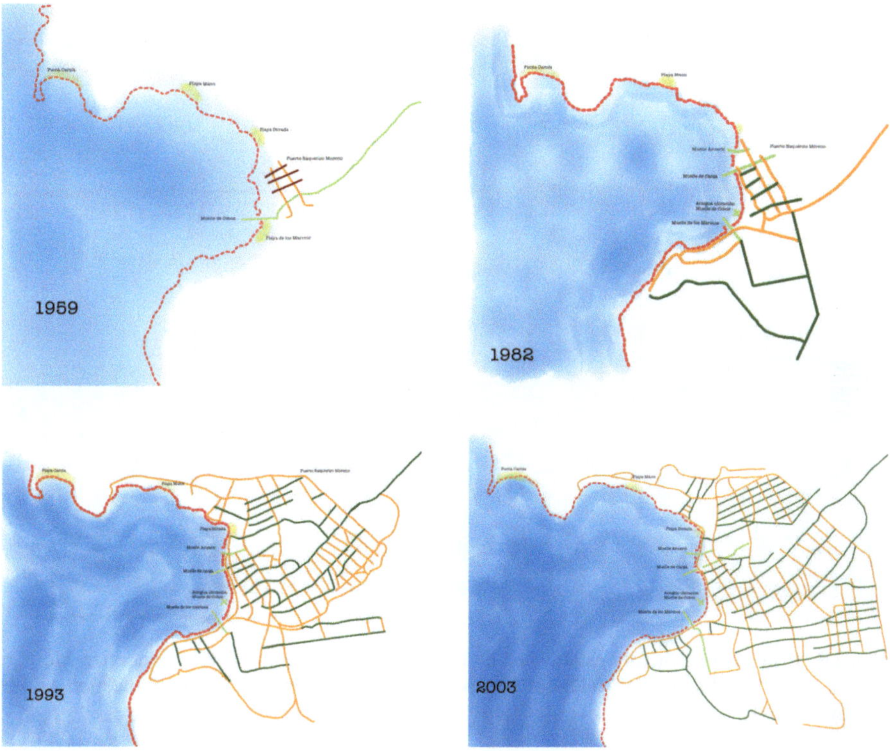

Fig. 2.7 Maps of Puerto Baquerizo Moreno, from aerial pictures taken in 1959, 1982, 1993, and 2003. (Maps: Antonella Segovia, John Dunn Insua)

In 1959, the settlement is composed of only a few houses and some ballasted roads (Fig. 2.7). Notably, the dock built by Manuel J. Cobos in Playa de los Marinos is in use at this time.

Then, in 1982, significant change in Puerto Baquerizo Moreno is already visible (Fig. 2.7). The built infrastructure has increased, with a greater number of streets and more blocks defined by them. Three docks have been constructed, including the one for the Ecuadorian Navy base, another for cargo management, and the "*Aquarius*," used exclusively by passenger boats and cruise ships during those years. Importantly, the actual airport started to be built at that time by the community, and a small hotel has been built beside Playa Mann; while an abandoned fish canning factory in the city outskirts, near Punta Carola (Fig. 2.7).

The aerial pictures from the Instituto Geográfico Militar also show how the density of the built-up form increases near the piers. Street infrastructure is also improved around them; in number and also in construction quality (Fig. 2.7).

The relevant changes demonstrated in the photographs from 1992 and 2003 involve urban growth (Fig. 2.7). The settlement has significantly increased the number of street blocks and roads. It has also expanded to Playa Mann, a new hub of

Fig. 2.8 Fish canning factory established by the Tous family, outside Puerto Baquerizo Moreno, known today as “La predial.” (Photo: Raymond Léveque. From the private collection of Mr. Whitman Cox)

infrastructural development where the Universidad San Francisco de Quito built a satellite campus.

Lastly, in photograph from 2007, the fourth and final pier in the town, “*Fishermen Dock*” is visible (Fig. 2.8). Together, these photos show the correlation between increased port infrastructure development—or crossing the permeable coastal membrane—and increased urban growth further inland.

With continued waterfront development came shifts in its management. In 2016, the Ecuadorian government changed the cargo management system from pallets to containers, in efforts to address issues with cost, efficiency, and food safety. However, this decision has been questioned by locals and media, as it was confirmed after three ships transporting goods from Guayaquil to Galápagos sunk within a nine-month period between their shipwrecks (Ministerio de Transporte y Obras Públicas 2015; Redacción Plan V, 2020).

The shift in cargo transport required significant adaptations to its infrastructure and logistics —some of which were never completed. Since then, barges are used to transport containers from the ships to the island, and their docking site became *La Predial*, near the former fish canning factory and beyond official city limits (Transporte de Carga hacia San Cristóbal, Galápagos 2015). Beyond logistical complexities, this move to La Predial, a private lot with no pier, allowed the dock that was formerly dedicated to cargo and fishing vessels—the second and longest of the four—to be remodeled as the primary pier for locals and tourist transport. In an

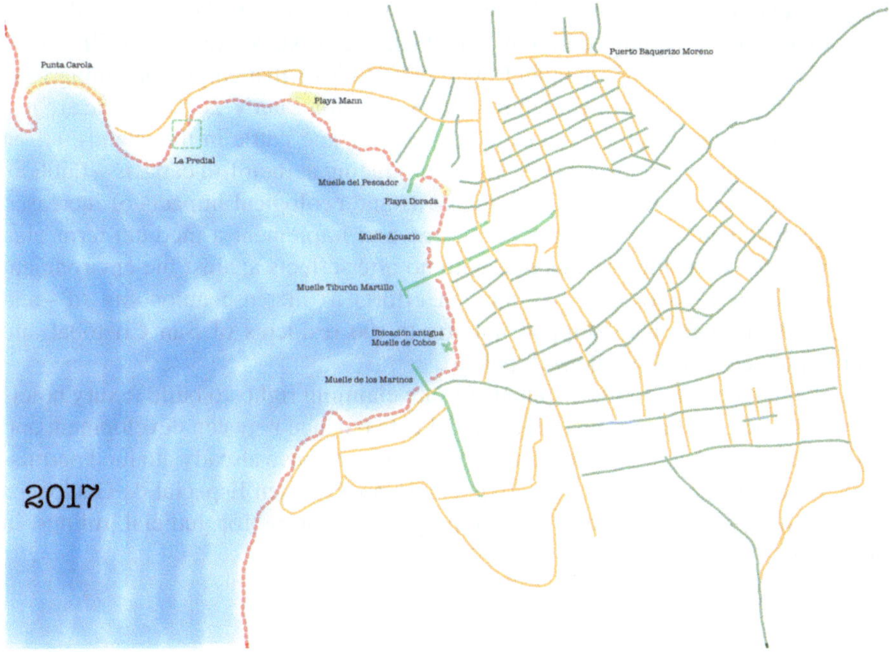

Fig. 2.9 Map based on aerial photographs from 2017 showing the Hammerhead Pier (center) and La Predial (upper left). (Graphic: Antonella Segovia and John Dunn)

aerial picture of Puerto Baquerizo Moreno from 2017, the remodeled dock is visible (Fig. 2.9). Because of its new T-shape, the dock was renamed as the "*Hammerhead Pier*."

The maps display how port infrastructure in Puerto Baquerizo Moreno has supported the consolidation and expansion of the urban core. From the 1980s to the early 2000s, the Hammerhead Pier and Aquarius Dock were landmarks that underpinned the town's center, hosting flocks of tourists and locals alike as the waterfront increased in built form. The Fishermen's Dock similarly reinforced the extension of Puerto Baquerizo Moreno's urban core.

However, as ports and surrounding areas have developed, community members—and local non-human residents who rely on the waterfront membrane—have faced new challenges. For example, since cargo is now being unloaded from *La Predial*, the resulting increase in heavy traffic and container-carrying trailers threatens the safety of marine iguanas' nesting habitats. This is only one conflict caused by the location of this port, nestled within the contested frontier between Puerto Baquerizo Moreno's urban zone and the protected Galápagos National Park.

It is worth highlighting, the consequences of using a private lot for all the cargo management implies a silent control over what comes in or out; or even worse, who are the ones that may be able to do it. This means a few people decides on a critical point of San Cristóbal and the Galapagos: its dependence on the food and goods that come from the Ecuadorean mainland. The sheer number of materials imported to the archipelago should prompt a thorough review of the assumed benefits of the new cargo management container system. The Galápagos Islands receive 6000 tons of cargo per month from the mainland, with over 44% of that cargo containing food that is produced outside of the archipelago (Redacción Plan V 2020). There are just over eight thousand hectares dedicated to farming and cattle in the islands, all on San Cristóbal, and even these lands are not producing at their full capacity (Público, 2021). Therefore, when considering the costs and ecological impacts of increased cargo management, it is necessary to design and implement a medium-term plan that reinforces local food production with no after-effects against the environment and establishes a food safety system, in efforts to increase autonomy of—and decrease disproportionate shipped imports to the residents of San Cristóbal and other populated islands in the archipelago.

In parallel, owners in this disputed area are claiming rights to build luxury hotels on their properties, further teetering the delicate balance between development and biodiversity conservation on the island. Some owners have already attained permissions in Ecuadorian courts (Consejo de Gobierno Regimen Especial de Galápagos 2015), while other cases are still in progress, fighting for judicial rulings in their favor.

Conclusions and Recommendations

Puerto Baquerizo Moreno is an interesting case study, which shows how port infrastructure creates a bridge to trespass the coastal membrane, and how urban development anchors to these points. This can be confirmed by the photographs analyzed in this research. They show increased urban growth alongside port infrastructural development in Puerto Baquerizo Moreno. These patterns also expose the risks of urban extension beyond the actual urban footrint, including how the port activities in *La Predial*, for example, impact the waterfront membrane: both local property owners and the world-renowned biodiversity of the Galápagos Islands.

The historical case study of port infrastructure in Puerto Baquerizo Moreno also uncovers further socio-ecological issues, including linkages between economic (in) efficiency and food security.

Plans have been announced for another redesign of cargo management from Guayaquil to the Galápagos Islands (Fig. 2.10). The Galápagos 2030 Plan, for example, proposes to establish one single port for the arrival of all materials from the mainland, as well as the distribution to all settlements in the archipelago (Plan Galápagos 2030 2012; Redacción Plan V 2020). However, this proposal has no implementation strategy in place at the time of writing. There have been unofficial plans to establish that main port in Baltra Island, in front of Santa Cruz north coast. Nevertheless, some studies have been started and later interrupted. In the meantime, it is necessary to conduct deeper analyses of how this redesign could impact the waterfront membrane and social-ecological systems of the archipelago. What are the environmental impacts of consolidating all imports to a specific island? How

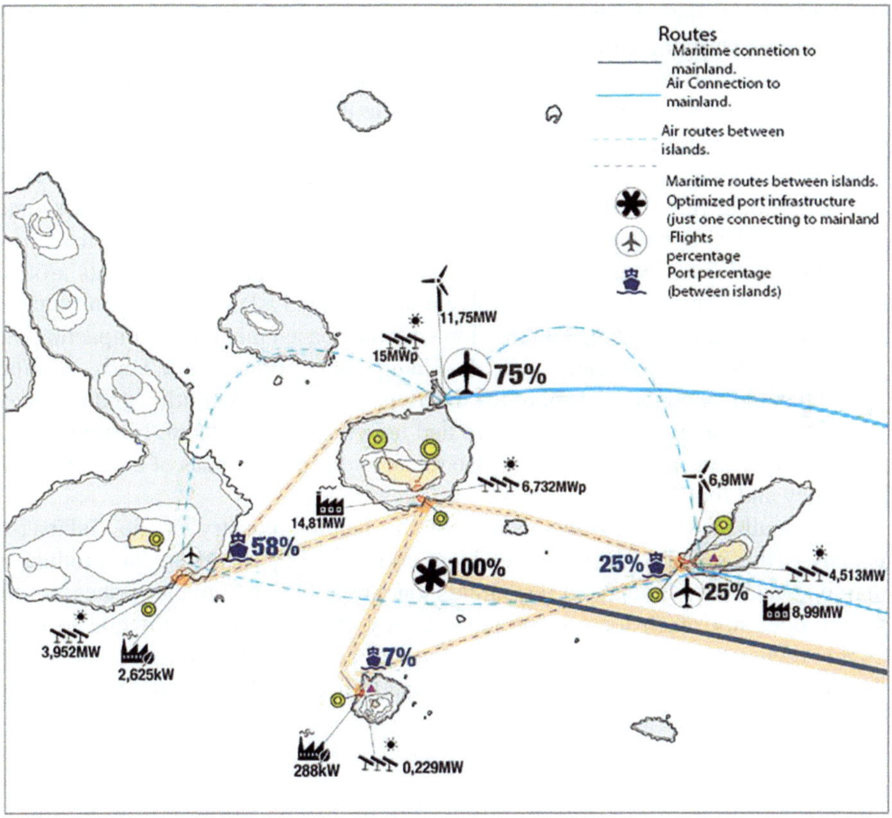

Fig. 2.10 Proposal for cargo management in the Galápagos Islands. (Graphic: Plan Galápagos 2030)

will local biodiversity and ecosystems respond? How will community members and local businesses on other inhabited islands be affected by the rearrangement?

Since the actual zoning of the marine reserve splits different zones and levels of conservation with more accuracy, through the recently developed annex of the 2030 Plan, we can establish more easily the areas that trend to preservation, scientific research, or artisanal fishing. Related to the argument at hand, currently, there is no designated way to control the inner boundaries inside the Galápagos Marine Reserve, especially the transition zones near ports. While the fluidity of ocean environments inherently presents challenges their governance, the current overlap suggests that massive shipping vehicles are regularly passing through "protected" waters, impacting the same water quality and species that the reserve was designed to protect. Therefore, the Galápagos Marine Reserve should include specific regulations and controls on human activities related to cargo shipping and maritime traffic between the archipelago's settlements and the mainland. As the Galápagos Marine Reserve encircles the entire archipelago, ships are unable to "go around it," but regulations may include marked boundaries between the reserve and port waters (e.g., specific shipping corridors and passageways).

The waterfront membrane is multi-dimensional, encompassing infrastructural, institutional, and social-ecological layers. While this chapter focused primarily on the impacts of infrastructural development, the waterfront membrane framework could be further expanded through dedicated research to the connections between the institutional (e.g., Galápagos Marine Reserve, Galápagos National Park) and social-ecological dynamics (e.g., tourism, fishing). For example, when considering "porosity" and live-aboard tourism by boat, many other permeable points around the archipelago emerge that even involve non-populated islands. This front implies more constant and widespread interaction with the coastal membrane, impacting the non-inhabited—but frequently visited—Galápagos National Park lands. Similar research could be conducted about the existing voids in the definition and management of the Galápagos Marine Reserve and how to approach them. Fishing activities have not been analyzed here but also relate to discussed issues of economic (in) efficiency and food security.

The small-scale or artisanal nature of the fishing industry in the Galápagos Islands is theoretically designed for local consumption. However, the product is mainly available to hotels and restaurants that are able to pay higher prices, so the supply does not reach local inhabitants. Lastly, it would be worthwhile to thoroughly review the ships that are responsible for connecting Galápagos Islands with the mainland, their characteristics and compatibilities with the port infrastructure existing over time.

References

Buque chino pasó de llevar pesca ilegal a cazar depredadores en Galápagos (2024, 23 July) https://www.primicias.ec/sociedad/buque-chino-pasa-pesca-ilegal-cazar-depredadores-galapagos-74746/

Carrión-Mero P, Sánchez-Zambrano E, Mata-Perelló J, Jaya-Montalvo M, Herrera-Franco G, Berrezueta E, Espinel RL, Baque M, Morante-Carballo F (2024) Geosites assessment in a volcanic hotspot environment and its impact on geotourism, Santa Cruz-Galapagos Islands, Ecuador. Int J Geoheritage Parks 12(1):147–167. https://doi.org/10.1016/j.ijgeop.2024.01.006

Consejo de Gobierno del Régimen Especial de Galápagos (n.d.) Plan de Desarrollo Sustentable y de Ordenamiento Territorial del Régimen Especial de Galápagos 2020–2030

Consejo de Gobierno Regimen Especial Galápagos, resolución 006-CGREG-11-II-2015 (2015)

Idrovo H (2013) Baltra – Base Beta: Galápagos y la Segunda Guerra Mundial. Fondo Editorial Ministerio de Cultura y Patrimonio

Jang J (2023) Chapter 1 – classification of membranes: with respect to pore size, material, and module type. In: Bui X, Guo W, Chiemchaisri C, Pandey A (eds) Current developments in biotechnology and bioengineering. Elsevier, pp 3–17

Latorre O (2011) Historia Humana de Galápagos (Segunda edición, 2014 ed.). Artes Gráficas Señal Impreseñal Cía Ltda

Moore R (2021) An encyclopedia of geography, history, and culture. ABC-CLIO, LLC

Plan Galápagos 2030 (2012). https://unidosporgalapagos.wordpress.com/wp-content/uploads/2021/05/plan-galacc81pagos-2030-2.pdf

Público P (2021, June 15) En San Cristóbal la agricultura aporta a la conservación. Periodismo Público. Retrieved Feb 27, 2025, from https://periodismopublicoec.com/2021/06/15/san-cristobal-agricultura-aporta-conservacion/

Redacción Plan V (2020) El largo y caro viaje de la carga que abastece a Galápagos – Plan V. https://Planv.Com.Ec/Historias/El-Largo-Y-Caro-Viaje-La-Carga-Que-Abastece-Galapagos/

Sarango S, Paucar SS (2017, August 21) El buque chino detenido en GalÃ¡pagos fue entregado a Inmobiliar. El Comercio https://www.elcomercio.com/actualidad/ecuador/buque-chino-detencion-galapagos-inmobiliar.html

Señalética para proteger nidos de iguanas marinas – Parque Nacional Galápagos (2016) Puerto Baquerizo Moreno; Galápagos, Ecuador. Retrieved Feb 27, 2025, from https://galapagos.gob.ec/senaletica-para-proteger-nidos-de-iguanas-marinas/

Transporte de Carga hacia San Cristóbal, Galápagos (2015) Pacific Cargo Line. Retrieved Feb 27, 2025, from https://pcl.ec/san-cristobal/

Veinte mil toneladas de productos fueron enviados a las islas Galápagos – Ministerio de Transporte y Obras Públicas (2015) Retrieved February 27, 2025, from https://www.obraspublicas.gob.ec/veinte-mil-toneladas-de-productos-fueron-enviados-a-las-islas-galapagos

Chapter 3
Embellishing Rocks: Accounts on the Cultural Imaginaries of the Galápagos Islands

Fernando Astudillo-Cueva, Miguel Andrés Guerra, Paul Rosero-Contreras, and Diego Quiroga

Abstract The Galápagos Islands are known worldwide for their unique ecosystems. The islands are often described as pristine and untouched by human activity, reinforcing an imaginary of the Galápagos as a "living laboratory" where humans can observe nature that is uninfluenced by man. However, human presence in the archipelago can be traced back to the mid-sixteenth century. Since the first recorded human arrival in 1535, representations of the islands have fluctuated according to both local and international geopolitics, varying from the "end of the world" to a "tourist treasure."

In this chapter, we present a recount of cultural imaginaries of the Galápagos Islands over time through a discourse analysis of selected materials (e.g., governmental documents, visitors' diaries, governmental reports, tourist brochures). We believe that on-island and off-island imaginaries of the Galápagos (re)shape local politics, demographics, social relationships, environmental policies, and cultural identities.

Keywords Galapagueños · Galápagos human history · Representation · Cultural identity · Ecotourism

---

F. Astudillo-Cueva (✉) · D. Quiroga
Anthropology, College of Social Sciences and Humanities, Universidad San Francisco de Quito (USFQ), Quito, Ecuador
e-mail: fjastudillo@usfq.edu.ec

M. A. Guerra
Civil Engineering, College of Sciences and Engineering, Universidad San Francisco de Quito (USFQ), Quito, Ecuador

P. Rosero-Contreras
Visual Arts, College of Communication and Contemporary Arts, Universidad San Francisco de Quito (USFQ), Quito, Ecuador

J. Dunn Insua et al. (eds.), *Human Galapagos*, Social and Ecological Interactions in the Galapagos Islands, https://doi.org/10.1007/978-3-032-08800-0_3

Introduction

Since the first recorded human arrival in 1535, the reasons why humans have valued and visited the Galápagos Islands have changed. Famous worldwide for their unique ecosystems, the islands are often contemporarily described as pristine and untouched by human activity, reinforcing the imaginary of the Galápagos as a living laboratory. However, since the mid-sixteenth century, representations of the islands have fluctuated according to both local and international geopolitics, varying from the "end of the world" to a "tourist treasure." These seemingly contradictory representations have been socially constructed over time, oscillating between descriptions of a long-term home or a place of temporary passage; a paradise or hell; a land of opportunity or a place of condemnation; islands enchanted by their beauty or by their curses that thwarted colonization.

Cultural imaginaries of the Galápagos, being products of the collective imagination, provide a glimpse into the diverse perspectives on the islands and the correspondingly complex political landscape (Hennessy 2018a, b; Hennessy and McCleary 2011; Villarreal 2010). In exploring the interplay of cultural imaginaries, representation, and the appropriation of nature, a strong conceptual grounding is essential. Here, we use the term cultural imaginary as the collective beliefs, values, and perceptions that shape how a society views the world around them. Representation delves into how these beliefs and perceptions are portrayed and depicted, whether through language, imagery, or other mediums (Anderson 2016; Strauss 2006). The appropriation of nature involves the ways in which people interact with and utilize natural environments as resources, often influenced by cultural constructs. By analyzing discourses around the cultural imaginary, representation, and appropriation of nature in the Galápagos Islands, we offer deeper insights into societal attitudes and behaviors towards the surrounding environments.

In this chapter, we propose eight dynamic cultural imaginaries of the Galápagos Islands over time: (1) The Enchanted Islands; (2) A Sea of Resources; (3) The Promised Land; (4) An Idyllic Escape; (5) A Strategic Stronghold; (6) The Living Laboratory; (7) A Tourist Treasure, and (8) Human Galápagos. We believe that internal and external imaginaries of the islands have (re)shaped local politics, demographics, social relationships, environmental policies, and cultural identities of the *Galapagueño* society.

To examine these complex dynamics, we use discourse analysis, which considers language and communication within specific contexts to unveil underlying ideologies, politics, power structures, and societal norms. A wide array of digitized sources informs our analysis, including governmental documents, scientific reports, academic articles, magazines, tourism advertisements, novels, and drawings. Government records provide insight into official policies, regulations, and narratives surrounding nature and territory. Scientific reports offer empirical data on environmental issues, shedding light on the intersection of science and culture. Magazines, tourist brochures, and advertisements present a commercialized representation of nature, often reflecting societal ideals and desires. Public-facing books and novels provide literary perspectives on nature, offering narratives that shape and reflect cultural values and beliefs. Drawings offer visual representations that may perpetuate or challenge cultural stereotypes and narratives about the environment.

The discourse analysis encompassed an initial search for key terms within digital and digitized documents related with the proposed imaginary categories, complemented by a thorough reading of relevant chapters or book sections. The selected and analyzed documents comprised published materials and secondary sources freely accessible through national and international digital repositories, including the National Archives of Ecuador, the UNESCO Digital Library, the Library and Archive of the Charles Darwin Foundation, the Eugenio Espejo National Library of Ecuador, the Aurelio Espinoza Pólit Ecuadorian Archives, the Allan Hancock Foundation Collection, and the Internet Archive.

The Enchanted Islands

Upon their discovery in 1535, the Galápagos were quickly shrouded in mystery and myth. Cultural imaginaries around the Galápagos began soon after human arrival, as the islands became the subject of fantastical tales and legends. While earlier inhabitation of the Galápagos by indigenous peoples of South America was proposed (Heyerdahl and Skjølsvold 1974), that hypothesis has since been challenged. More recent evidence suggests a later arrival of humans to the archipelago (Anderson et al. 2016; Sanchez and Freire 1990), which positions the first cultural imaginaries of the Galápagos in the sixteenth century.

Fray Tomás de Berlanga, during a voyage from Panama to Perú, was the first person to document the existence of the Galápagos. In his letter to the Majesty of Spain, Berlanga depicted the islands as a harsh landscape and described one island–possibly Floreana–as a worthless place outside the limits of normality (de Berlanga 1884). In his words:

> …but on the whole island I do not think that there is a place where one might sow a bushel of corn, because most of it is full of very big stones, so much so that it seems as though at some time God had rained stones upon the land; and the earth that there is, is like slag, worthless… (Figueras Vallés 2009; Vargas 1986).

The discovery of the Galápagos archipelago ushered in a new era of exploration. Early explorers encountered what they deemed to be strange and otherworldly creatures, leading to the islands' reputation as a land of monsters and mythical beasts. These tales laid the foundation for an early cultural imaginary spread by sailors, whalers, and pirates who traversed the remote waters: the Galápagos as enchanted islands.

During the sixteenth century, as explorers began visiting the Galápagos as a stopover on their voyages, cartographers published the first maps of the islands. The archipelago was initially displayed on Abraham Orteloius' 1570 "Theatrum Orbis Terrarum" (Map of the Terrestrial Globe), generally considered the first modern atlas, and Gerardus Mercator's 1569 "Nova et Aucta Orbis Terrae Descriptio ad Usum Navigantium Emendate Accommodata" (New and Improved Description of the Earth for the Use of Navigators), known for its use of the Mercator projection. However, the precise location of the Galápagos was uncertain, and early legends surrounding the unclear positions or "movement" of the islands were spreading. For example, the seasonal mist that covers the islands between June and November annually hid the islands from Spanish authorities, making them disappear into the middle of the ocean. Accordingly, the nickname of "Enchanted Islands" grew in popularity. As Civallero points out,

> The name alluded to the difficulty of Spanish pilots and captains to place the islands on their sea charts. From their medieval perspective, they were difficult to find because they were "enchanted", i.e. put under a charm that made them change position over and over again, or disappear and appear at will (Civallero 2022, pp. 14–15).

During this period, the Galápagos were merely enchanted or haunted islands on the little-known maritime frontier of the Pacific, then called the South Sea, and colonization was not a priority. To the Spanish Crown, these mysterious, remote, and wealthless islands on the frontier of their empire were just another nuisance (Guerra 2018, p. 471).

By the seventeenth century, pirates had begun visiting the Galápagos during routes on the Pacific. The lack of Spanish presence on the islands and little control over this remote territory, combined with the cultural imaginary of enchantment, created the perfect scenario for a pirate hideout in the South Sea. The Galápagos served as a stopover and refilling station for collecting water, replenishing food, and recovering energy. Retellings of renowned visitors, such as William Ambrose Cowley or William Dampier, created impressions of the Galápagos as a pirate settlement. These stories and others about hidden treasures throughout the archipelago continue to this day, frequently mentioned by tour guides. Contemporarily, tours to sites including the Cave of Pirates on Floreana Island reinforce the nostalgia of the enchanted islands (Fig. 3.1).

The mythical creatures, apparent curses that prevented the islands' colonization, and legends of hidden treasure continued to shape the rhetoric and history of the archipelago. Descriptions about the Galápagos as non-suitable for humans included legends of shipwrecks, deaths, crimes, and other tragic endings for those who dared to attempt to inhabit the islands. For instance, aboard the whaler *Acushnet* in 1841, a young Herman Melville wrote about the Galápagos, "In no world but a fallen one could such a land exist" (quoted in Otterman, 1993, p. 2; Guerra 2018, p. 476). These representations and the difficulty to properly map the archipelago could explain the lack of interest by the Spanish to colonize the archipelago (Bassett 2009; Guerra 2018; Latorre 2011).

The imaginary of the Galápagos as enchanted islands was perpetuated in Western literature by Herman Melville's (1854) work, who described the archipelago as desolate, haunted, inhospitable, the place of shipwrecks, exile and suffering; presenting a dramatic description of its geography and justifying the lack of humans living on the islands.

Fig. 3.1 Construction of the caves in Floreana Island are attributed to pirates passing by the archipelago during the eighteenth century. Stills from the video: Dark Paradise: Humans in Galápagos. (Rosero-Contreras 2019)

> Another feature in these isles is their emphatic uninhabitableness. It is deemed a fit type of all-forsaken overthrow that the jackal should den in the wastes of weedy Babylon, but the Encantadas refuse to harbor even the outcasts of the beasts. Man and wolf alike disown them. Little but reptile life is here found: tortoises, lizards, immense spiders, snakes, and that strangest anomaly of outlandish nature, the iguana. No voice, no low, no howl is heard; the chief sound of life here is a hiss. On most of the isles where vegetation is found at all, it is more ungrateful than the blankness of Atacama. Tangled thickets of wiry bushes, without fruit and without a name, springing up among deep fissures of calcined rock and treacherously masking them, or a parched growth of distorted cactus trees (Melville 2002, p. 6).

A Sea of Resources

During the 17th and 18th centuries, the Galápagos Islands became a focal point for appropriation of nature and extraction. The natural resources of the islands, including exotic wildlife, attracted international explorers seeking to exploit the archipelago for economic gain. This early relationship to nature resulted in the establishment of seasonal settlements, natural resource exploitation, and introduction of non-native species (Froyd et al. 2010). A second imaginary emerged: a sea of wild resources to extract.

At this time, Galápagos giant tortoise meat was in high demand on the international market. Collecting tortoises was a profitable business dominated by canning companies in the U.S. and U.K., but the exotism and the highly reputable meat flavor created a demand in foreign capitals. As reptiles that can decrease their body temperature and metabolic rate, Galápagos giant tortoises can survive up to a year without food or water. In a world before freezers, this was a godsend (Nicholls 2021, p. 49). Explorers' testimonies of tortoises surviving long trips on boat decks even created stories of this reptile as a source of protection from scurvy during exploration in the Pacific (Morrell 1832; Porter 1822). In a sense, these creatures were the refrigerators that sailors needed so dearly (Conrad and Pastron 2014, p. 23).

An interesting parallel with whaling elsewhere was the popularization of Galápagos giant tortoise soup and steaks during the California Gold Rush. In the wild western frontier, this exotic meat was highly esteemed, advertised as the finest food for the "lovers of good eating," and provided a sense of comfort after highly dangerous mining activities (Conrad and Pastron 2014). Historical records indicate that during the early nineteenth century, ships collected up to 300 individual Galápagos tortoises on every trip, thousands of individuals were exported by the 1860s (Conrad et al. 2020; Townsend 1925). While earlier imaginaries had depicted the islands as enchanted, haunted, and not fit for humans, the archipelago was later seen as a sea of resources to exploit, and soon after, to inhabit (Figs. 3.2 and 3.3).

The Promised Land

The Galápagos Islands were claimed by the independent nation of Ecuador in 1832, deemed to be a strategic location for continued resource extraction and emergent agricultural production. During this period, the islands became a frontier of the Ecuadorian state and attracted colonization efforts. Colonization brought a new imaginary to the Galápagos: a promised land for capitalism.

During the early phases of colonization in the Galápagos, land was considered free and available for development, according to the interest of the modern nation. The Ecuadorian government incentivized the colonization of the Galápagos by offering exploitation concessions to private companies, tax reductions or subsidies, and exercising minimal control over local activities. During the first decade of

Fig. 3.2 A live Galápagos tortoise exhibited at the New York Zoological Park in the late 1890s. This image was originally published by Charles H. Townsend in *The Galapagos Tortoises in Their Relation to the Whaling Industry* (Townsend 1925). The caption reads: "A Galápagos tortoise (*Testudo vicina*) from Albemarle Island, which has tlived for twenty years in the New York Zoological Park. Weight when received: 140 pounds. Present weight: 305 pounds" (Townsend 1925, p. 97)

colonization, the state's presence was minimal. For settlers and developers, this presented the unique possibility of expanding their businesses in a remote location away from government oversight. In this regard, the Galápagos became a land of opportunity.

The colonization of the Galápagos Islands involved the development of permanent agricultural settlements, which soon functioned as *de facto* penal colonies, and infrastructure to support resource extraction activities funded by private companies (Astudillo and Jamieson 2021). Companies such as Companía Colonizadora del Archipelago, owned by Jose de Villamil, or Hacienda El Progreso, owned by Manuel J. Cobos and Jose Monroy, imagined the Galápagos as an outpost for exporting goods, and even contraband (Stahl et al. 2020). Examples of early agricultural colonies and monocrop plantations include the Asylum of Peace, Hacienda El Progreso, and Hacienda Santo Tomás. Convicts were sent to labor there under harsh conditions, further contributing to the appropriation of nature and transformation of the physical and social landscapes (Fig. 3.4).

Settlers and visitors to the islands in the late nineteenth century described the difficulties of early development. Charles Mann discussed the harsh realities of colonization on San Cristóbal Island -formerly known as Chatham, as follows:

Fig. 3.3 Live Galápagos tortoises taken by sailors. Stills from the film To Tropic Seas aboard Velero III. Allan Hancock Foundation part of Smithsonian Institution Archives, Hancock Pacific-Galapagos Expeditions (1933–1935). (Photography by W. Chas Swett)

> …From the hides, orchil, and tortoise oil [Cobos] formed this farm, valued to-day at £3o,ooo sterling, for before his time there was only a small patch cultivated by a negro recluse. Having cajoled a few labourers to accompany him, he held them afterwards by force, treating them with harshness and cruelty, and paying them a minimum wage, consisting of vulcanite tokens supposed to represent coin, and as he was the owner of the only vessel trading to the island, no one could escape... (Mann 1909, p. 34).

This new imaginary of the islands as a place of prosperity–for some–started to surface in the private sector, slowly replacing earlier notions of the cursed archipelago (Carbo 1894; Martínez 1915; Tous 1933; Vidal Gormáz 1890).

However, many new immigrants viewed the Galápagos as a place of isolation and suffering. The government of Ecuador used the islands as sites of criminal exile and relocated convicts to the islands to work for private interests. The relocation of convicts to work on plantations created an imaginary of the archipelago as dangerous and isolated. In addition, the limited government presence and plantation owners' self-declarations as emperors perpetuated the imaginary of abandonment from the state (Astudillo and Jamieson 2021; Latorre 1991). The social landscape in the early colonies was so volatile that it spurred two rebellions on San Cristóbal and Floreana, where plantation owners were killed by their imprisoned workers

Fig. 3.4 These images provide a rare view of the entire sugar mill infrastructure and the sugar cane fields of El Progreso Plantation circa 1905. The plantation produced and exported refined sugar to the Ecuadorian mainland. The industrial agricultural activities at El Progreso started in the 1860s, resulted in the permanent human occupation of the Galápagos (Top: Bognoly and Espinosa 1905, p. 137; bottom: Archivo Histórico del Guayas http://fotografiapatrimonial.gob.ec/web/es/galeria/element/1223)

(Bognoly and Espinosa 1905). During the trial of Manuel J. Cobo's assassination, life on the islands is violently described as follows:

> The people of Chatham, or rather the slaves of the inhuman Manuel J. Cobos, tired of tolerating so many abuses and working for years and years without any remuneration, and only receiving punishments, tortures and whips, we resolved not to continue under the pressure in which we were living... All the inhabitants of Galápagos know that Cobos shot five individuals; that six were whipped to death; that he banished 15 men to the deserted islands and of these some died of hunger; it is well known that Cobos whipped without mercy and one day he applied this infamy to an unfortunate woman named Rosa Agustina. The authorities never punished crimes of this kind, because they were all under the dominion of Cobos, the king of Galápagos, the tyrant of Ecuador... this gave rise to all the slaves, or rather the majority, unanimously agreed to put an end to this tiresome and tyrannical life and a plot was organized (Webster 1904, pp. 113–114).

Throughout the late nineteenth century, societal tensions were high on the inhabited islands of the Galápagos, and multiple cultural imaginaries co-existed: both a promised land for capitalism and a place of exile for prisoners.

It is worth noting that at the same time, naturalists and field scientists started to record the native flora and fauna of the archipelago. The published reports from scientific expeditions, mainly from the U.S., created a scientific interest and opened the door for the later imaginary of the Galápagos as a living laboratory (Banning 1933; Chubb 1933; McBride 1918; Stewart 1911; Tanner 1888, 1893).

An Idyllic Escape

Moving into the 1930s, the Galápagos were considered a frontier not fully colonized and also the official exile place of Ecuador. However, notions of nature conservation and preservation were growing in importance worldwide. Reports by scientists like George Banning, a member of the Smithsonian-funded Hancock expeditions to the Galápagos, observed human impacts and threats to endemic fauna:

> Such hazard [the environmental] to the hapless tortoise, however, was all but negligible as compared to the scourge that followed the coming of the white man early in the sixteenth century. Then it was that two or three processes of real extermination began... when these edible reptiles were carried away by the thousands, many thousands have since been taken, not only for food but for oil, while such important animals as the dog, hog and cat, now thriving in wild state throughout much of the old tortoise country, have well night done for the rest through the destruction of eggs and young. But for these imported animals, introduced by the whalers and by the various ephemeral colonist from Ecuador, it is likely that the Charles tortoise (*Testudo galapagoensis*) would still be with us; and it is possible, even despite them, that he is. (Banning 1933, p. 5).

Given the uniqueness, remoteness, and now vulnerability of the islands, Ecuador began taking steps to protect the "pristine" nature of Galápagos. Concurrently, war in Europe triggered international migration to the islands and the next cultural imaginary: an idyllic escape from global conflict.

Galápagos: World's End, an account from renowned naturalist William Beebe (1924), suggested a fascinating and appealing character of the archipelago, far from earlier representations of danger and hostility (Fig. 3.5). In the preface, Beebe writes:

Fig. 24
Head of a Giant Marine Iguana
Slightly enlarged
From a painting by Isabel Cooper

Fig. 3.5 William Beebe's work showcased the fascinating and unique native fauna and flora of the Galápagos, countering earlier descriptions of the species as wild and dangerous. His efforts sparked the early wave of scientific tourism to the islands in the early twentieth century (Beebe 1924)

> This book is in no way intended as a technical contribution to our present knowledge of the fauna and flora of the Galápagos Archipelago. It is a record of spontaneous observation, a crystallization of the more obvious and characteristic impressions of the land and the sea, of birds, reptiles, fish and insects, whose forms and colors and personalities made us wish that every one of the six thousand minutes of our stay could be lengthened to hours or days (Beebe 1924).

The trend of focusing on the beauty and idealism of the archipelago was also followed by Ecuadorian artists Manuel Rendon and Paulette de Rendon (1978). Escape from western society also motivated the travels of Dr. Richard and Dore Strauch, a German couple who moved to Floreana island (formerly Charles Island) to follow a theosophist treatise based on isolation from the world, vegetarianism, and contact with nature. Similarly, the Whitmer family also ascribed to the idea that nature would heal their ill son and chose Floreana as their permanent home.

During the same time period, notions of escapist, idyllic ecotourism ventures started budding in the U.S. and Europe. One of these dreamers was Eloise Bosquet de Wagner, known by the press as the "Empress of the Galápagos Islands" or "the Baroness" (Markey 2011a). The Baroness considered the archipelago to be a promising destination for luxury travelers, and when she traveled to the islands in the company of two romantic partners, she fueled a media frenzy:

> …The trio landed on Charles Island with a vague notion of opening a resort, and soon were the subject of bizarre stories circulating in the American press. These attracted the attention of American millionaires who made Charles Island a stop-off on their yacht cruises, delighted at a chance to gawk at genuine free-love advocates living in tropical sin… (Markey 2011a).

The imaginary of the Galápagos as the nudist archipelago was internationally recognized at the time. Even scientists identified their explorations to the Galápagos as trips to "nudist islands." However, as Waldo Schmitt, a Smithsonian scientist and member of the Hancock expeditions, argued: nudism would not be a permanent trend in the archipelago, because they went without clothes for practical reasons—not philosophical ones. In other words, they weren't nudists; they were just naked. In his words, "The nudism so widely touted in the papers is more or less a figment of the imagination. We have a picture of the only nudist on the island—the Wittmer infant…" (Waldo Schmitt Papers, Record Unit 7312, box 87. In: (Markey 2011b) (Fig. 3.6).

Drama and conflict among settlers on Floreana Island during the 1930s ended in the mysterious disappearance and death of former colonist Dr. Ritter, the Baroness, and one of her romantic partners. The story has been highlighted in books and documentaries that portray the Galápagos as a place of suffering, lack of authority, and murder (Javier et al. 2015; Wittmer 2013). Amidst the imaginary of idyllic escape, another that coexisted simultaneously was an ungoverned archipelago forgotten by the state, an imaginary that is still sometimes used as a form of protest (Castillo de Vargas et al. 2005; Hennessy 2018b). Until the late 1930s, most foreign projects of relocation to the Galápagos had collapsed after drama, violence, or rebellions, perpetuating the idea of the islands' curses against humans and complicating the history of cultural imaginaries over time (Latorre et al. 1990).

Fig. 3.6 Top: Allan Hancock, an American explorer, took this photograph of Dr. Friedrich Ritter (left) and Dore Strauch (right) sitting in their open-air home, "Friedo," on Floreana Island in January 1932. The history of this couple was popularized by European media, depicting them as residents of a "Garden of Eden" nudist colony in the Galápagos, which contributed to the imagery of the islands as an idyllic landscape (Allan Hancock Foundation Collection, https://scalar.usc.edu/works/the-allan-hancock-foundation-archive/media/ritter%2D%2Dstrauch-friedo-floreana-galpagos-1932). Bottom: Heinz and Margret Wittmer with their older son Harry and baby, Rolf Hans, 1934. Waldo Schmitt Papers, RU 7312, box 89. Image #2011–1470

A Strategic Stronghold

World War II heavily influenced another imaginary: the Galápagos as a strategic location to protect military interests. The archipelago re-emerged as a critical place to defend the U.S. and one of its most precious investments, the Panama Canal.

Early in 1940, the General Board of the Navy and the Army-Navy Joint Board reached the conclusion that preparations must be made to operate constant air patrols over a wide area southwest of Panama. They recommended that patrol squadrons of seaplanes be based near Guayaquil on the Ecuadorian coast, in the Gulf of Fonseca in Nicaragua, and in the Galápagos. The Galápagos, it was decided, were to be the key installations, and they were subsequently fortified by both the U.S. Army and Navy (Department of the Navy 2018, pp. 3–4). In this context, the U.S. Army and Navy built a base on Baltra Island with a lease from the Ecuadorian government (Galarza Zavala et al. 2016; Idrovo 2013; Steinsleger 1986). From the islands, 600 miles from the mainland, naval seaplanes flew northwest to Corinto, Nicaragua and east to Salinas, Ecuador. Army landplanes assisted in covering the southern route. The Galápagos became the focal point for a wide arc of aerial patrols protecting western approaches to the Panama Canal.

During the same decade, the opening of the Isabela Penal Colony marked the Ecuadorian government's final attempt to exile political convicts to the archipelago (Constante 2003; Durán Arias 1998; Rodas Ziadé and Vivanco 2012). According to Tamayo (Tamayo Rivera 2018), establishing this prison in the Galápagos also responded to contemporaneous, international debates on models of prisoner insertion. The idea was to colonize otherwise inhospitable areas and, at the same time, strategically reinsert the prisoner into a "productive" regime. In the case of Isabela, the extreme isolation and lack of escape options justified the creation of the penal colony. Once again, this ended in a rebellion due to harsh living conditions and human rights violations, strengthening the older imaginaries of the Galápagos as a place inhospitable for humans.

The Living Laboratory

In 1959, international preservation and conservation interests inspired the creation of the Charles Darwin Foundation. On July fourth of the same year, Ecuadorian government responded to these interests establishing the Galápagos National Park, and in 1979, the archipelago was declared a UNESCO World Heritage Site. These recognitions, along with over a century of scientific expeditions since Charles Darwin explored the islands as a naturalist himself, created the next imaginary: the living laboratory.

The Galápagos had long been considered an ideal research site. Since the late nineteenth century, scientists and other visitors had described the archipelago as nearly untouched by humans due to its isolation, making it a natural "living laboratory" for observing evolutionary processes (Hennessy and McCleary 2011). Darwin's legacy was undoubtedly pivotal in establishing the Galápagos as an epicenter for biologists across the world (Mariscal. 1969). Accordingly, foreign evolutionary biologists were among the first to advocate for conservation of the archipelago and largely responsible for the establishment of GNP and Charles Darwin Research Station (Bowman 1960). The original mission of the Charles Darwin Foundation follows: "CDF was founded under the auspices of UNESCO and IUCN to provide the highest quality scientific research, technical advice, and practical expertise on conserving and protecting the archipelago. We have since acted as the GNPD's primary partner and the source of key research and capacity-building to direct the Park's conservation efforts for the sustainable development of Galapagos and its local community (https://www.darwinfoundation.org). On the other side, the foundation act of the Galápagos National Park mentions that "considered the high ecological value of this archipelago and declared 97% of the land area a National Park, making it the first protected area in the country. The decree declared "national reserve parks under the exclusive domain of the State, for the preservation of flora and fauna, the lands that form the islands of the Colon or Galápagos archipelago, except for the lands owned to date by the colonists of the archipelago and those that had already been legally owned by the State" (Executive decree N-17, July fourth 1959). This demonstrated increasing desires to conserve endemic flora and fauna, as well as honor the birthplace of evolutionary thought–preserving a site of both pristine nature and scientific history (Hennessy and McCleary 2011) (Fig. 3.7).

Conservation research and related tourism became central pillars of the region's economic development (Romanova et al. 2013). However, tensions arose between the two under the subconscious belief that humans were separate from–and inherently incompatible with–nature (Denevan 1992). Following the designation of the national park, local populations were seen as a negative force to be constrained in order to protect the natural wonders that made the region famous (de Groot 1983). From conservationists' perspectives, growth, and development (e.g., increasing local and visitor populations, coastal infrastructure, energy consumption) threatened the sustainability of the living laboratory and increased its vulnerability to climate impacts (Tapia et al. 2008). Maintaining the unique ecological and evolutionary processes of the Galápagos depended on some degree of isolation that had characterized the archipelago for most of its history. Local communities made efforts to preserve the islands, primarily governed by the regulations of the national park, but tourism was–and is–many residents' main source of income. Revenue from this booming tourism industry led to further population growth, increased resource consumption, and economic inequality on the islands, causing ecosystems to shift from their earlier status as "an almost untouched place" (Kerr et al. 2004). Therefore, tensions existed between conservation scientists' research agendas and local residents' economic interests, and the archipelago was–and remains–immersed in a profound socio-ecological crisis.

Fig. 3.7 The cover of *LIFE* magazine in 1958 portrayed the Galápagos Islands as a wildlife treasure. The composite image was created by photographer Alfred Eisenstaedt and artist Rudi Freund, both of whom were sent by *LIFE* magazine during the 1957 UNESCO Mission—a three-month expedition to the archipelago aimed at documenting its endangered fauna. Such initiatives, designed to attract international attention, contributed to the establishment of the Charles Darwin Foundation and the Galápagos National Park in 1959. (Image: https://www.etsy.com/listing/1577841490/september-8-1958-life-magazine-galapagos)

A Tourist Treasure

The most pressing issue currently facing the Galápagos is the balancing of conservation goals, tourism, and corresponding development. In the 1960s–70s, tourism led to booming island economies, as cruise ship fleets expanded and towns in the Galápagos began to offer land-based options (Brewington 2013). Tourism quickly became residents' main source of income, and the number of visitors to the islands has increased dramatically since: from 11,765 in 1979 to 185,028 in 2013 (Villacis and Carrillo 2013). Currently, the Galápagos Islands welcome more than 250,000 tourists per year, creating yet another imaginary: Ecuador's tourist treasure.

To promote tourism in the Galápagos, the government has made efforts to preserve "pristine" land, but released a management plan in which the local community and nature remain at odds. The perception of untouched ecosystems upholds a worldview in which humans are separate from nature, as residents are removed from specific environments to avoid contaminating pure nature with an "unnatural humanity" (Adams and Hutton 2007). To keep these islands intact, residents are forced to maintain landscapes that tourists desire, while the needs and desires of growing local populations are overshadowed (Mathis and Rose 2016). It is these policies and regulations that create a disconnect between communities and their surrounding environments. Though biodiversity conservation is critical, it must also be understood that humans and their multiple worldviews are an integral part of nature itself (Sanderson et al. 2002).

The expanding tourism industry is closely intertwined with conservation strategies, local development, and governmental regulations in the archipelago. A greater number of tourists and locals require more infrastructure, which involves increased resource extraction and, consequently, greater environmental impacts. Simultaneously, increased tourism generates revenue for scientific research, conservation efforts, and community development projects (Mathis and Rose 2016). Further, local residents are subject to governmental regulations in parks and reserves, centered around ensuring the pristine and primitive landscapes advertised to tourists. The GNP restricts residential practices and imposed protections for a number of species, which has changed how locals interact with the environment (Mathis and Rose 2016). For example, activities such as raising livestock, agriculture, and fishing are limited, impacting the once-thriving subsistence agricultural and fishing industries (Duffy 2006; Grenier 2007). Consequently, farmers and fishers consider these rules to be unjust (Mathis and Rose 2016).

Local communities in the Galápagos have emphasized that tourism and conservation initiatives can cause environmental injustices (Vaccaro et al. 2013). While residents have been involved in tourism since its beginnings, they have not received its resulting benefits equitably. Often, tourism revenue is allocated right back to tourism-related development projects, despite locals lacking basic services (Mathis and Rose 2016). For example, many residents lack potable water, adequate healthcare, and sufficient educational systems (Brewington 2013). This external control of money and power leads to uneven development within a host community. Further, conservation programs typically do not provide adequate support to local inhabitants (Campbell 2007). The asymmetric distribution of power between inhabitants and institutions has generated resentment amongst locals against the national park (Mathis and Rose 2016). In conversations with San Cristóbal residents, they mention that the community lacks a sense of ownership towards the islands and feel invaded since their home became a tourist destination. For Galapagueños, the designation of GNP has not been advantageous, and despite adapting their lifestyles to required environmental regulations, they have not received adequate compensation. However, these tensions are typically obscured by portrayals in nature documentaries and travel brochures, where the archipelago is simply a tourist's paradise (Hennessy and McCleary 2011).

In response, the tourism industry, conservation organizations, and local governments have made efforts to increase local communities' participation and representation. Understanding that community participation plays a key role in social resilience and empowerment, industries have developed training programs, so residents can work as tour guides, accommodation operators, and service providers. Additionally, they have encouraged the creation of local businesses and the sale of handicrafts (Mathis and Rose 2016). For example, many fishers have transitioned their livelihoods to hosting snorkeling trips for tourists. Relatedly, tour operators will often buy live animals (e.g., seahorses, octopuses) from other fishers to strategically place them in locations where they subsequently bring tourists, creating memorable experiences for tourists while financially supporting local community members (Quiroga 2013).

Though the COVID-19 pandemic severely impacted all aspects of life in the Galápagos, following the height of the pandemic, the government has been determined to revitalize the islands' imaginary as a tourist treasure. San Cristóbal, the administrative capital of the Galápagos and the second-largest population center of the archipelago, has begun marketing itself as the Capital of Paradise. This new branding strategy is now placed above way finding maps, in illuminated kiosks installed along the urbanized waterfront of Puerto Baquerizo Moreno, and in other touristic areas. The logo is not observable in the non-touristic blocks of the city.

This social media project named "The Capital of Paradise, San Cristobal Galápagos", is a set of promotional videos together with Instagram and Facebook posts created by the tourism board of San Cristóbal, portrays an imaginary of a carefree tourist treasure. Tropical beaches, surfing, snorkeling, bicycling, hiking, running, and wildlife are the major features of the video, with the majority of screen time occupied by white tourists. These examples support how the Galápagos, in the aftermath of the pandemic, promoted a paradisiacal imaginary of islands and beaches, reinforcing its status as a world-class tourist destination (Fig. 3.8).

Fig. 3.8 In 2021, the municipality of San Cristóbal launched a touristic campaign declaring the islands as the Capital of Paradise, reinforcing the imaginary of a unique tourist treasure for foreign visitors. Watch the YouTube videos here: https://www.youtube.com/@thecapitalofparadisesancri3441

Final Thoughts: Human Galápagos

In this chapter, we identified several cultural imaginaries that have (re)shaped local politics, demographics, social relationships, environmental policies, and cultural identities in the Galápagos Islands from the 16th to the 21st centuries. The legacy of these periods continues to influence the contemporary socio-economic landscape of the Galápagos, highlighting the complex interplay between geopolitics, tourism, resource extraction, culture, and environmental conservation in the modern era.

We observe the emergence of a new imaginary in the second half of the twenty-first century, one that integrates the agendas of conservation, tourism, and the well-being of the archipelago's inhabitants. In the late twentieth century, many *Galapagueños* temporarily migrated to the mainland to pursue undergraduate and graduate degrees in key fields such as law, medicine, engineering, agriculture, public policy, tourism, or conservation. Since the early 2000s, both Ecuadorian and international universities have established campuses and research stations in the archipelago, offering higher education to locals. This expansion in educational opportunities is expected to reduce the demand for professionals traveling from the mainland to provide technical and professional services and will shape future conservation, research, and tourism agendas in alignment with the interests of local communities.

In August 2024, after 26 years, the local government of the Galápagos increased the entrance fees to protected areas for both international and national tourists. This decision aligns with the objectives of promoting a tourism model that supports ongoing conservation efforts, generates benefits for local citizens, and helps stabilize the number of tourists arriving in the province (Consejo de Gobierno del Régimen Especial de Galápagos, 2024). This evolving imaginary positions people as integral agents within the Galápagos, moving away from extreme conservationist movements that have questioned human occupation of the archipelago. We believe that the Human Galápagos imaginary acknowledges the consolidation of a distinct local society in the Galápagos, where concepts of identity, tradition, economy, food security, arts, and cultural heritage play vital roles, challenging and disregarding the idealistic and romantic imaginaries historically associated with the islands.

References

Adams WM, Hutton J (2007) People, parks and poverty: political ecology and biodiversity conservation. Conserv Soc 5(2):147

Anderson B (2016) Imagined communities: reflections on the origin and spread of nationalism (Revised edition). Verso

Anderson A, Stothert K, Martinsson-Wallin H, Wallin P, Flett I, Haberle S, Heijnis H, Rhodes E (2016) Reconsidering Precolumbian human colonization in the Galápagos Islands, republic of Ecuador. Lat Am Antiq 27(2):169–183. https://doi.org/10.7183/1045-6635.27.2.169

Astudillo FJ, Jamieson RW (2021) The Galápagos as penal colony: exile, peonage, and state control at the hacienda El Progreso, 1878–1904. Punishment Soc 25(1):42–59. https://doi.org/10.1177/14624745211013100

Banning GH (1933) Hancock Expedition of 1933 to the Galapagos Islands (No. 10; Bulletin of the Zoological Society of San Diego). Zoological Society of San Diego

Bassett CA (2009) Galapagos at the crossroads: Pirates, biologists, tourists, and creationists battle for Darwin's cradle of evolution. National Geographic Books. http://books.google.ca/books?hl=en&lr=&id=QcgXzcCWMIYC&oi=fnd&pg=PA8&dq=galapagos+pristine+myth&ots=0_T6Cw3NGZ&sig=xbHzofpaeZL1UTUMQywhmYHlsd0

Beebe W (1924) Galapagos: world's end. G. P. Putnam's Sons

Bognoly JA, Espinosa JM (1905) Las islas Encantadas: Ó, El archipiélago de Colón. Impr. y lit. del Comercio

Bowman (1960) A biological reconnaissance of the Galapagos Islands during 1957. United Nations Educational, Scientific and Cultural Organization, 1–56

Brewington L (2013) The double bind of tourism in Galapagos society. In: Walsh SJ, Mena CF (eds) Science and conservation in the Galapagos Islands: frameworks & perspectives. Springer, pp 105–125. https://doi.org/10.1007/978-1-4614-5794-7_6

Campbell LM (2007) Local conservation practice and global discourse: a political ecology of sea turtle conservation. Ann Assoc Am Geogr 97(2):313–334. https://doi.org/10.1111/j.1467-8306.2007.00538.x

Carbo LF (1894) El Ecuador en Chicago. [A.E. Chasmar y cia.]

Castillo de Vargas B, Herrera C, Jarrín H, Ospina P (2005) Desde las islas encantadas: Historias de vida de colonos en Galápagos. Corp. Editora Nacional

Chubb LJ (1933) Geology of Galapagos, Cocos, and Easter islands, Honolulu, Hawaii: http://hdl.handle.net/2027/mdp.39015023272050

Civallero E (2022) A history of Galapagos in 15 documents: from Ortelius' map to Darwin's diary. Charles Darwin Foundation

Conrad C, Pastron A (2014) Galapagos tortoises and sea turtles in gold rush-era California. Calif Hist 91(2):20–39. https://doi.org/10.1525/ch.2014.91.2.20

Conrad C, Garwood N, Gibbs J (2020) Digitized Galapagos Tortoise Whaling Data from 1831–1868. *Anthropology Datasets*. https://digitalrepository.unm.edu/anthro_data/9

Constante A (2003) Basalto: Etapa de Terror y Lágrimas Durante la Colonia Penal en Isabela. Gráficas Pato

de Berlanga T (1884) Letter to his majesty ... Describing his voyage from Panamá to Puerto Viejo. In: Pacheco JF, de Cárdenas F, Torres de Mendoza L (eds) Coleccion de Documentos Ineditos relativos al Descubrimiento, Conquista y Organizacion de las Antiguas Posesiones Españolas de América y Oceania.Tomo XLI, Cuaderno II. Imprenta de Manuel G. Hernandez

de Groot RS (1983) Tourism and conservation in the Galapagos Islands. Biol Conserv 26(4):291–300. https://doi.org/10.1016/0006-3207(83)90093-9

Rendón PE de (1978) Galápagos: Las últimas islas encantadas. Casa de la Cultura Ecuatoriana

Denevan WM (1992) The pristine myth: the landscape of the Americas in 1492. Ann Assoc Am Geogr 82(3):369–385. https://doi.org/10.1111/j.1467-8306.1992.tb01965.x

Department of the Navy (2018) Building the Navy's Bases, vol. 2 (part III). The Advance Bases Chapter XVIII Bases in South America and the Caribbean Area, Including Bermuda. http://

public1.nhhcaws.local/content/history/nhhc/research/library/online-reading-room/title-list-alphabetically/b/building-the-navys-bases/building-the-navys-bases-vol-2.html
Duffy R (2006) The politics of ecotourism and the developing world. J Ecotour 5(1–2):1–6. https://doi.org/10.1080/14724040608668443
Durán Arias J (1998) La colonia penal. El muro de las lágrimas. sin editor
Figueras Vallés E (2009) Domingo de Pasión en un paraíso: Fray Tomás de Berlanga en las Islas Galápagos (1535). Archivo Dominicano: Anuario 30:251–286
Froyd CA, Lee JA, Anderson AJ, Haberle SG, Gasson PE, Willis KJ (2010) Historic fuel wood use in the Galápagos Islands: identification of charred remains. Veg Hist Archaeobotany 19(3):207–217. https://doi.org/10.1007/s00334-010-0239-1
Galarza Zavala J et al (2016) Bases militares norteamericanas en el mundo: Capítulo especial Ecuador. Ministerio de Relaciones Exteriores y Movilidad Humana
Grenier C (2007) Conservación contra natura: Las islas Galápagos. Quito, EC: Universidad Andina Simón Bolívar, Sede Ecuador/AbyaYala /Instituto Francés de Estudios Andinos, IFEA/ Embajada de Francia en Ecuador/Institut de Recherche Pour le Développement, IRD. http://repositorio.uasb.edu.ec/handle/10644/7420
Guerra S (2018) Las Galápagos: Piratas y mapas de las islas en los márgenes del imperio. Colon Lat Am Rev 27(4):469–489. https://doi.org/10.1080/10609164.2018.1560139
Hennessy E (2018a) The Galápagos Islands: a natural laboratory? Arcadia. https://arcadia.ub.uni-muenchen.de/arcadia/article/view/170
Hennessy E (2018b) The politics of a natural laboratory: claiming territory and governing life in the Galápagos Islands. Soc Stud Sci. https://doi.org/10.1177/0306312718788179
Hennessy E, McCleary AL (2011) Nature's Eden? The production and effects of 'pristine' nature in the Galápagos Islands. Island Stud J 6(2):131–156
Heyerdahl T, Skjølsvold A (1974) Archaeological evidence of pre-Spanish visits to the Galápagos Islands. Kraus Reprint Co.
Idrovo H (2013) Baltra-Base Beta: Galápagos y la Segunda Guerra Mundial. Fondo Editorial, Ministerio de Cultura del Ecuador
Javier A, Pedro C, Esteban RB (2015) Floreana: Islamundo en Galápagos. Editorial Abya – Yala
Kerr S, Cardenas S, Hendy J (2004) Migration and the Environment in the Galapagos: An analysis of economic and policy incentives driving migration, potential impacts from migration control, and potential policies to reduce migration pressure. Motu Economic and Public Research, Working Papers 03_17, Article 03_17. https://ideas.repec.org//p/mtu/wpaper/03_17.html
Latorre O (1991) Manuel J. Cobos. Emperador de Galápagos. Fundación Charles Darwin para las Islas Galápagos
Latorre O (2011) Historia Humana de Galápagos. Nuevas Investigaciones. Academia Nacional de Historia
Latorre O, Netherly P, Dickinson de Salomón J, Cummins K (1990) The curse of the giant tortoise: Tragedies, mysteries and crimes in the Galapagos Islands. Libri Mundi
Mann A (1909) Yachting on the Pacific: together with notes on travel in Peru, and an account of the peoples and products of Ecuador. Duckworth & Co.
Mariscal. (1969) Charles Darwin and conservation in the Galápagos islands. Biol Conserv 2(1):44–46. https://doi.org/10.1016/0006-3207(69)90114-1
Markey M (2011a, July 12) The Empress of the Galapagos Islands, Part 2. Smithsonian Institution Archives. https://siarchives.si.edu/blog/empress-galapagos-islands-part-2
Markey M (2011b, August 9) The Empress of the Galapagos Islands, Part 3. Smithsonian Institution Archives. https://siarchives.si.edu/blog/empress-galapagos-islands-part-3
Martínez NG (1915) Impresiones de un viaje. Talleres de policía nacional
Mathis A, Rose J (2016) Balancing tourism, conservation, and development: a political ecology of ecotourism on the Galapagos Islands. J Ecotour 15(1):64–77. https://doi.org/10.1080/14724049.2015.1131283
McBride GM (1918) The Galapagos Islands. Geogr Rev 6(3):229–239. https://doi.org/10.2307/207816

Melville H (2002) The enchanted isles. Hesperus, London. http://archive.org/details/enchantedisles0000melv
Morrell B (1832) A narrative of four voyages. J. & J. Harper
Nicholls H (2021) Human perceptions of Galapagos tortoises through history. In: Gibbs JP, Cayot LJ, Aguilera WT (eds) Galapagos Giant Tortoises. Academic, pp 49–62. https://doi.org/10.1016/B978-0-12-817554-5.00007-1
Porter D (1822) Journal of a cruise made to the Pacific Ocean. Wiley & Halsted
Quiroga D (2013) Changing views of the Galapagos. In: Walsh SJ, Mena CF (eds) Science and conservation in the Galapagos Islands. Springer, New York, pp 23–48
Rodas Ziadé FP, Vivanco K (2012) Galápagos, Prisión de Basalto: Terror y lágrimas en la Isla Isabela (1946–1959) (1ra. ed). Fondo Editorial Ministerio de Cultura del Ecuador
Romanova A, Yakushenkov S, Lebedeva I (2013) Media coverage of cultural heritage and consumerism in modern society. https://www.semanticscholar.org/paper/Media-Coverage-of-Cultural-Heritage-and-Consumerism-Romanova-Yakushenkov/e1b66cd852aa064f4a0c7cae16288f033273c43f
Rosero-Contreras P (2019) Dark Paradise: humans in Galápagos. Experimental Film. Dos Islas Studio, TAtchers Art Management and Universidad San Francisco de Quito
Sanchez A, Freire AM (1990) ¿Arqueología en Galápagos? ARAS (Boletín Arqueológico) 1:28–31
Sanderson EW, Jaiteh M, Levy MA, Redford KH, Wannebo AV, Woolmer G (2002) The human footprint and the last of the wild: the human footprint is a global map of human influence on the land surface, which suggests that human beings are stewards of nature, whether we like it or not. Bioscience 52(10):891–904
Stahl PW, Astudillo FJ, Jamieson RW, Quiroga D, Delgado F (2020) Historical ecology and archaeology in the Galápagos Islands: a legacy of human occupation, 1st edn. University Press of Florida
Steinsleger J (1986) Bases militares en América Latina. El Conejo
Stewart A (1911) A Botanical survey of the Galapagos Islands (Vol. 1). Proceedings of the California Academy of Sciences. http://hdl.handle.net/2027/hvd.32044106365349
Strauss C (2006) The imaginary. Anthropol Theory 6(3):322–344. https://doi.org/10.1177/1463499606066891
Tamayo Rivera MF (2018) La Colonia Penal en la Isla Isabela. In INEHPOL Instituto de Estudios Históricos de la Policía Nacional (Vol. 20, pp. 47–64). Studio 21
Tanner ZL (1888) Galapagos Islands; Extract from report of Capt. Tanner, 1887–88; extract from report of Capt. Tanner, 1887–88. Ernst Mayr Library of the Museum of Comparative Zoology, Harvard University. https://curiosity.lib.harvard.edu/expeditions-and-discoveries/catalog/38-990054634040203941
Tanner ZL (1893) Report upon the investigations of the U.S. fish commission steamer albatross from July 1, 1889, to June 30, 1891
Tapia W, Ospina P, Quiroga D, Reck G, González JA, Montes C, Cruz E, Milstead B, Wolff M, Watkins G, Carrión C, Piu M, Oviedo M (2008) Toward a shared vision of Galapagos: The archipelago as a socioecological system. In Galápagos Report 2007–2008. CDF, GNP & INGALA
Tous L (1933) Mi explicación sobre la Sociedad Nacional de Galápagos. Artes Gráfica "Senefelder"
Townsend CH (1925) The Galapagos tortoises in their relation to the whaling industry – a study of old logbooks. Zool Sci Contrib N Y Zool Soc 4(3):55–135. https://doi.org/10.5962/p.203768
Vaccaro I, Beltran O, Paquet PA (2013) Political ecology and conservation policies: some theoretical genealogies. J Polit Ecol 20(1). https://doi.org/10.2458/v20i1.21748
Vargas JM (1986) Fray Tomas de Berlanga y el descubrimiento de las islas Galápagos. In: Boletín de la Academia Nacional de Historia, vol LXVIII. Academia Nacional de Historia, pp 37–59
Vidal Gormáz F (1890) El archipiélago de las Galápagos. Imp. Nacional
Villacis B, Carrillo D (2013) The socioeconomic paradox of Galapagos. In: Walsh SJ, Mena CF (eds) Science and conservation in the Galapagos Islands: frameworks & perspectives. Springer, pp 69–85. https://doi.org/10.1007/978-1-4614-5794-7_4

Villarreal P (2010) Representaciones e imaginarios del entorno en Galápagos: Un recorrido a las islas desde distintas miradas [masterThesis, Universidad Andina Simón Bolívar, Sede Ecuador]. http://repositorio.uasb.edu.ec/handle/10644/2326

Webster EA (1904) Los crímenes de Galápagos (Archipiélago de Colón), Asesinato de Valdizán; asesinato de Cobos y Reina. Imprenta "El Telégrafo"

Wittmer M (2013) Floreana: a woman's Pilgrimage to the Galapagos (Revised edition). Beaufort Books

Chapter 4
Endemic Architecture. The Evolution of Architectural Forms in the Galapagos Islands: The Case of San Cristóbal Waterfront Restaurants

Cristina Vargas Guarderas, Jaime López-Andrade, and Elise O'Brien

Abstract The plants and animals of the Galápagos Islands have evolved distinct phylogenetic traits that are found nowhere else on Earth. Each island is home to species with unique characteristics not seen on any of the neighboring islands, reinforcing their biological isolation. These island-specific organisms are known as endemic species. As Charles Darwin noted in *On the Origin of Species (1859)*, an endemic species is shaped by the particular conditions of the place it inhabits and cannot be found elsewhere. A parallel can be drawn in the field of architecture. Similarly, the study of typologies is a way to classify buildings that share similar features in their form. Analogous to evolution in flora and fauna, the development of building typologies is also influenced by the context in which they evolve. In this sense, typologies that originate in a specific location and are not found elsewhere may be considered architecturally endemic.

This chapter presents a morphological analysis of the urbanized waterfront in Puerto Baquerizo Moreno, located on San Cristóbal Island, with a special focus on restaurant building types. Restaurants in San Cristobal exhibit typological characteristics unseen on mainland Ecuador that have emerged through adaptations driven by landscape, usage, and culture. These unique characteristics of the locale have resulted in newly emergent typologies of the Galapagos that have evolved in dialogue with the environment.

C. V. Guarderas (✉)
Institute of Architecture and Urban Form (AFORU), Universidad San Francisco de Quito (USFQ), Quito, Ecuador
e-mail: cmvargas@usfq.edu.ec

J. López-Andrade
Institute of Architecture and Urban Form (AFORU) and Instituto de Energía y Materiales, Universidad San Francisco de Quito (USFQ), Quito, Ecuador

E. O'Brien
Department of Landscape Architecture, University of Oregon, Eugene, OR, USA

J. Dunn Insua et al. (eds.), *Human Galapagos*, Social and Ecological Interactions in the Galapagos Islands, https://doi.org/10.1007/978-3-032-08800-0_4

The methodology employed juxtaposes and compares the urban and architectural scales of the built waterfront across specific historical periods. This chapter focuses on the typology of the restaurant, which has historically served fishermen, naval personnel, and tourists at different points in time. It examines how the morphology of this building typology reflects social and demographic shifts on the island of San Cristobal, in this way shaping the built environment. The research analyzes building typological transformations as adaptive responses to environmental, social, and geographical conditions, using San Cristóbal's waterfront as a case study.

Galapagos Colonization and the Evolution of San Cristóbal's Built Environment

To understand the urban growth of the Galápagos Archipelago, it is essential to review the historical arrival and settlement of its inhabitants. Key events throughout history have been documented chronologically through maps, drawings, and written accounts from different visitors to the Archipelago. The first documented European landing occurred in 1535, when Spanish bishop Fray Tomás de Berlanga was blown off course during a voyage between Panama and Ecuador. Around 1684, the *Batchelor's Delight*, a British pirate ship, became one of the first known pirate vessels to dock in the archipelago. During the eighteenth century, many visitors were linked to the Golden Age of Piracy. The Galápagos Islands were historically described as extremely difficult to access. The harsh climate, rugged geography, and lack of fresh water were among the most frequently cited challenges in early documents, making the islands an inhospitable environment for human habitation. Few remnants remain from the pirate era, but one of the most notable is Post Office Bay, a site that functioned as an informal mail system. Pirates would leave letters there to be collected and delivered by other passing sailors, enabling long-distance communication with families back home.

In 1830, José de Villamil was appointed General Governor of the Galápagos Archipelago, and colonization efforts began on Floreana Island with the arrival of a group of soldiers. In 1832, the Ecuadorian government formally endorsed this initiative, and under the administration of President Juan José Flores, the Galápagos were officially claimed as Ecuadorian territory. The first official settlement, named "Asilo de Paz," was established in the highlands of Floreana Island. However, the extreme natural conditions made life nearly impossible for early settlers, and many perished during these colonization attempts. In 1837, Villamil resigned and was succeeded by the British officer James Williams. His brutal rule, combined with the harsh environment, led to a rebellion by the settlers. In 1845, a small group of survivors relocated to Wreck Bay on San Cristóbal Island. The islands' early history is marked by repeated colonization attempts and a high mortality rate caused by the unforgiving conditions of the archipelago. In 1861, during the presidency of Dr.

Gabriel García Moreno, the Galápagos Islands were officially established as the first island province of Ecuador, with San Cristóbal designated as its capital.

In 1866, the *Empresa Industrial de Orchilla y Pesca*, owned by Manuel J. Cobos and José Monroy, was established to promote colonization on San Cristóbal Island. One of its major initiatives was the creation of Hacienda El Progreso, which aimed to ensure a stable food supply. Worker and farmer settlements began to form around the hacienda. Over time, however, laborers grew resentful of the inhumane treatment and extreme poverty imposed by Cobos. In 1904, this discontent culminated in a rebellion, during which Cobos was killed by his own workers. Hacienda El Progreso is a landmark in Galápagos history, marking the start of long-term human settlement in the archipelago. Today, several remaining structures provide insight into life during the early colonial period. Records suggest that the main house of the hacienda resembled traditional hacienda architecture from mainland Ecuador, while the workers' dwellings were comparable to the huasipungo houses of the Andean highlands—modest, single-story units with limited openings and steeply pitched thatched roofs, often occupying nearly half the height of the structure. These homes were arranged around communal open spaces, forming an organic village layout likely dictated by the landowners of the estate.

Historical records also indicate that the buildings were elevated above ground level, following the natural topography. This design choice created a ventilated space between the damp soil and the living areas, helping to reduce interior humidity. The main house, in contrast to the smaller dwellings, was a two-story structure. Its second floor included large, covered terraces at both the front and rear, as shown in period photographs. Although Hacienda El Progreso was located in the humid zone of San Cristóbal Island—a cool region with high humidity and fertile, organic soils—the inclusion of terraces appears more typical of buildings in warmer climates. However, given the main house's position atop the plantation, these terraces likely functioned as lookout points over the cultivated land.

The rest of the house followed design strategies suited to the cool climate of the region, including small, controlled openings, predominantly closed façades, and steeply sloping roofs. The overall architectural composition demonstrated a hybrid of styles, blending elements of traditional Andean hacienda architecture with influences from Ecuador's coastal regions.

In the twentieth century, two major events influenced the arrival of visitors and settlers to the Galápagos Archipelago: the inauguration of the Panama Canal and the onset of World War I. Following several failed negotiations with the Ecuadorian government to rent or purchase the islands, the United States maintained interest in the archipelago due to its strategic location for naval surveillance. In 1919, the U.S. Navy began conducting military and naval exercises between southern Panama and the Galápagos. Later, in 1938, President Franklin D. Roosevelt visited the islands, reaffirming U.S. geopolitical interest and establishing diplomatic relations aimed at protecting both the Panama Canal and the broader Americas. In 1940, amid rumors from residents of German submarine sightings in Ecuadorian waters, the Ecuadorian government authorized the United States to establish Beta Base on Baltra Island—a small island near the geometric center of the archipelago. This location provided a strategic advantage by enabling access to all surrounding islands and facilitating triangulation with Corinto in Nicaragua and Salinas on the Ecuadorian mainland, where the U.S. maintained additional military bases (Idrovo 2005).

In 1942, construction began on a variety of facilities at Beta Base, including office buildings, hangars, movie theaters, beer pubs, and housing. Despite the site's remote location, far from the continental United States, the base was planned with an urban layout and architectural style that reflected its American origins. Most buildings were constructed using prefabricated wood and lava stone. Today, the ruins of the officers' club remain as tangible evidence of the architectural planning and spatial logic of these wartime Galápagos structures.

Beyond the development of Beta Base on Baltra Island, the American military also constructed infrastructure on other islands in the archipelago. On San Cristóbal, they built an aqueduct to carry water from the highlands down to the port, and on Isabela Island, they established a military station. However, the most impactful legacy of American military construction was the network of airstrips. These airstrips—still in use today—serve as essential links between the Galápagos Islands and mainland Ecuador, facilitating both civilian travel and supply transport.

During this period, the civilian population of the archipelago increased significantly, growing from approximately 810 to 3200 inhabitants. American ships and airplanes became a common sight in the surrounding seas and skies. One of the greatest challenges for residents was securing a reliable supply of fresh water. The highlands of San Cristóbal, particularly the area near Wreck Bay, became a key source from which barges transported water to Baltra. This vital freshwater resource was a major factor behind the migration of settlers to San Cristóbal and led to the construction of the first land-based airport in the Galápagos. The American Rock base was officially decommissioned after World War II, in 1946 (Idrovo 2005).

After the American occupation ended, the Ecuadorian government permitted residents to dismantle remaining structures on Baltra and repurpose the materials for construction. The salvaged timber became widely known as "Baltra Wood." Using these resources, many homes and buildings were erected along the coasts of Santa Cruz and San Cristóbal. Some of these structures remained standing until only a few years ago, though most have disappeared due to lack of maintenance. These buildings were typically one or two stories tall, characterized by compact floor plans, small windows, and modest balconies facing the street. In several cases, the ground floor was built of stone, while the upper floor was constructed from wood.

Prior to 1950, the Galápagos Islands experienced minimal external influence due to their isolation, lack of urban and architectural infrastructure, and the absence of a land-based airport. Located 906 km from the mainland, early residents had to become self-sufficient, relying on local natural resources for survival. Hunting, gathering, and farming were essential practices for sustaining life in the archipelago (Stahl et al. 2020). In the decades that followed, the introduction of daily passenger flights to Baltra serving Santa Cruz and San Cristóbal, along with weekly cargo shipments from mainland Ecuador, gradually improved accessibility. These developments slowly transformed the remote archipelago into a significant tourist destination.

Regulations established by the Galápagos National Park limit human habitation to just 3% of the total land area, with the remaining 97% designated as protected parkland. Of the 13 major islands in the archipelago, only five—Santa Cruz, Baltra, San Cristóbal, Floreana, and Isabela—host permanent human settlements and infrastructure. Among these, Santa Cruz and San Cristóbal feature the most developed urban centers. Today, approximately 85% of the Galápagos' resident population is concentrated in the port towns of these islands (INGALA 2005). The town of Puerto Ayora on Santa Cruz, for instance, is known for its vibrant nightlife, numerous restaurants, and extensive tourism infrastructure.

San Cristóbal began its own process of urbanization, as evidenced in the growth patterns shown in the maps below (Fig. 4.1). These maps reveal increased density and expansion inland—both toward the highlands and in the direction of the town of El Progreso. Landscape urbanists refer to this type of transformation as urban morphogenesis—a process in which building is understood as a cultural and social act that generates complex, evolving spatial patterns (Kropf 2017).

"Nothing could be less inviting than the first appearance," wrote Charles Darwin about San Cristóbal in his historical accounts. Known at the time as Chatham Island, San Cristóbal had an estimated population of about 150 inhabitants by 1879 (Stahl et al. 2020). New architectural and infrastructural development began around 1941 with the arrival of the military base, which was later converted into a commercial airport in 1953. Urbanization accelerated during the 1950s, particularly along the main waterfront and extending inland toward the rural area of El Progreso. At that time, there were no souvenir shops or restaurants catering to foreign tourists.

Human settlers often chose locations that had already been used in previous colonization attempts, building upon earlier settlements (Stahl et al. 2020). Because access to fresh water was the primary concern, initial settlement occurred in the highlands. It was not until the establishment of the naval base in Puerto Baquerizo Moreno that development along the waterfront began in earnest. One of the first types of businesses to appear were restaurants, which commonly served the sailors stationed or visiting the island.

Today, it is common for Galapagueños to rent out their urban homes in Puerto Ayora and Puerto Baquerizo Moreno and instead reside on rural properties in the highlands (Kvan and Karakiewicz 2018). This trend may signal future urbanization pressures in the currently rural zones of both Santa Cruz and San Cristóbal. Ecotourism, with its fluctuating annual population of approximately 250,000 visitors, is expected to continue exerting development pressure on these areas (Granda et al. 2013).

At present, the architecture of San Cristóbal—and the Galápagos more broadly—has been described by several scholars as closely resembling that of continental Ecuador. This is largely because the population of the islands consists primarily of migrants or descendants of migrants (López 2024). However, a closer examination reveals that the forms of many buildings reflect adaptations more closely tied to climate and functional use than to the cultural backgrounds of their inhabitants.

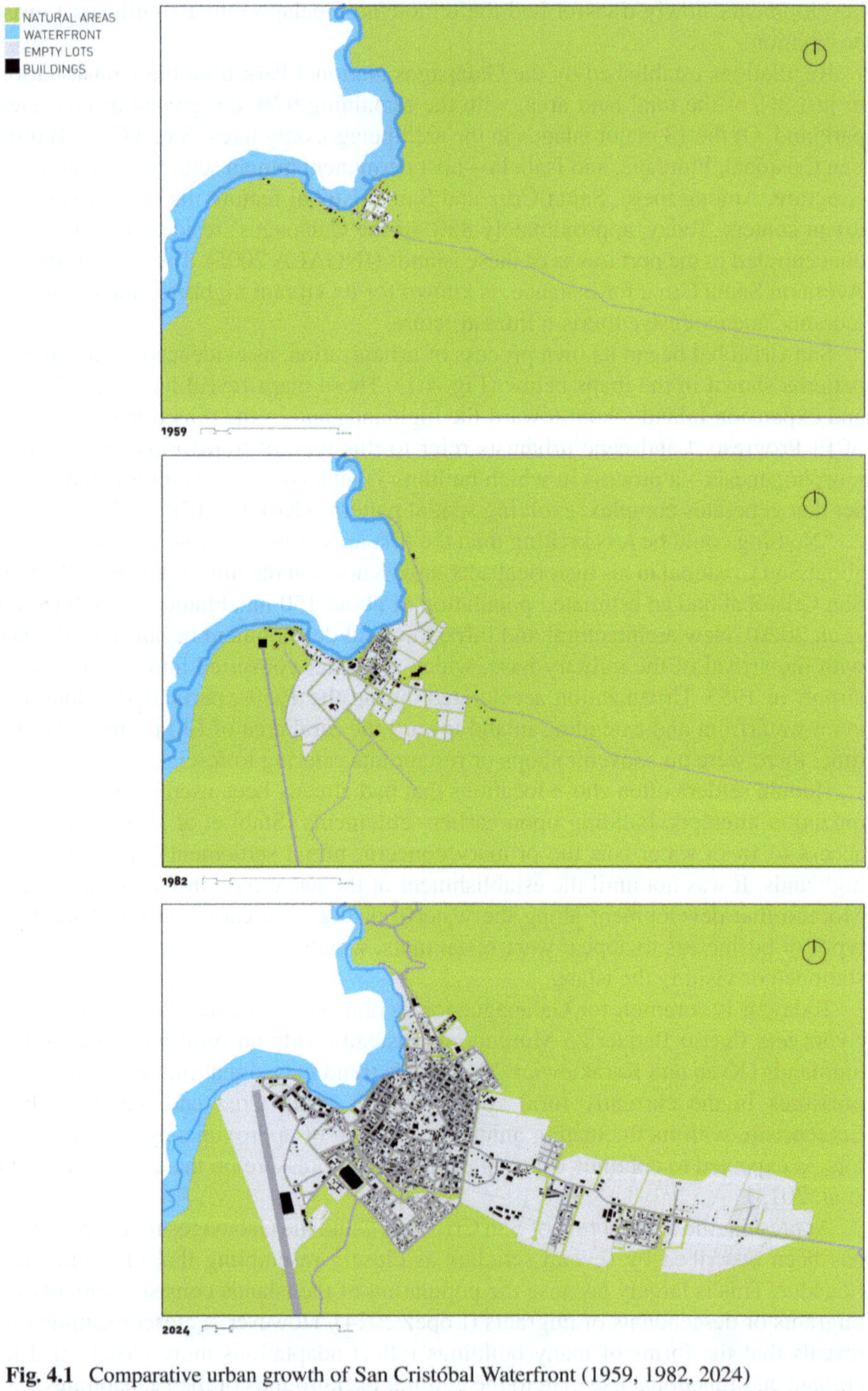

Fig. 4.1 Comparative urban growth of San Cristóbal Waterfront (1959, 1982, 2024)
Note. Visualization of Puerto Baquerizo Moreno's waterfront area showing spatial transformations and land-use changes across three key years: 1959, 1982, and 2024. The figure highlights the development of buildings, persistence of natural areas, and emergence of empty lots over time

This research examines the morphology of Puerto Baquerizo Moreno's urbanized waterfront—a place that was once considered wild and nearly uninhabitable. Today, it accommodates a variety of tourism infrastructures, including luxury hotels, high-end restaurants, souvenir shops, and other visitor-oriented facilities.

The Analogy Between Darwin's Theory and the Concept of Endemic Architecture

Charles Darwin's foundational theory of evolution was published in 1859 in *On the Origin of Species by Means of Natural Selection, or the Preservation of Favored Races in the Struggle for Life*. Today, his theory continues to serve as a cornerstone in numerous scientific fields, including biology, computational modeling, and even design (Losos et al. 2013). In this work, Darwin introduced natural selection as the mechanism by which evolution occurs—a process through which favorable variations are retained over time, allowing for the emergence of highly adapted forms. As Huxley (1944) explains, natural selection offers a mechanistic account of life's apparent purposiveness. This evolutionary process establishes a relationship between organic form and the geographic conditions in which it develops. In isolated environments such as volcanic islands, these conditions give rise to specific and limited forms (Darwin 1859).

Today, sustainable architecture emphasizes the importance of responding to natural conditions—including sun, wind, rain, soil type, and locally available materials—all of which ultimately shape architectural form. In contrast, vernacular architectural traditions are typically imagined as evolving across centuries, with their origins lost in deep history. In *Architecture Without Architects*, Rudofsky (1965) challenges conventional definitions of architecture by highlighting the ingenuity of traditional, non-professional builders. He emphasizes the global diversity of architectural forms and celebrates how ordinary people construct buildings that respond to local climate, needs, and culture.

The Galápagos Islands, however, do not possess such a prehistoric architectural tradition. With no indigenous population, the settlements of San Cristóbal—and the population of Galapagueños—have emerged only within the past few centuries. The urban history of the Galápagos can be traced back only a few decades. As a result, the islands have no vernacular architectural heritage; most buildings are spontaneous constructions reflecting the basic needs and intuitive choices of their residents, rather than the product of an evolved architectural lineage or technical discourse on sustainability. While some traditions were transplanted from mainland Ecuador—especially the coastal Guayas region—Galápagos has always been shaped by a diverse influx of international visitors and settlers. It remains, in many ways, a global colony (Sánchez 2022). In this context, where "place" is defined by human occupation and meaning, Galápagos functioned for much of its history as a non-place.

Architectural forms or building types were initially brought to the Galápagos by settlers, and over time, they have begun to evolve their own distinctive traits through continuous interaction with the San Cristóbal context. Building types that were present during the early phases of settlement are now rapidly transforming, shaped by the pressures of tourism and the challenges posed by its local context.

Urban morphologists such as Christopher Alexander and Michael Batty have explored the forces shaping the development of the built environment. In *Notes on the Synthesis of Form* (Alexander 1964), Alexander rejects the notion of an organic evolution in architectural form, emphasizing instead the primacy of social structures. Nevertheless, architectural practices often reflect adaptive behaviors analogous to those found in natural systems. In the context of this study, the restaurant typology in San Cristóbal exemplifies such responsiveness, adapting to the unique environmental and socio-political conditions of the archipelago. This typology emerged early in the development of Puerto Baquerizo Moreno, where fishermen began establishing restaurants to serve the growing naval population—one of the first waves of immigrants to the island. These early restaurants adopted the vernacular architectural traditions of Guayaquil and were constructed using wood left behind by the American Army and Marines at Base Beta on Baltra Island after it was turned over to Ecuador in 1946. The United States had originally sought a 99-year lease of the Galápagos Archipelago, but the proposal was rejected by the Ecuadorian public. When American forces withdrew, they left behind a stockpile of materials, which were repurposed to build early waterfront structures in San Cristóbal, including several of these restaurants.

This chapter examines the notion that, much like organic forms, architectural forms in the Galápagos Archipelago evolve in response to the islands' unique geographical and ecological conditions. In most regions, traditional architecture is shaped by vernacular styles and the use of locally sourced materials. In contrast, the Galápagos lacks these defining features due to the absence of an indigenous cultural tradition and strict environmental regulations governing construction practices (López 2024). As a designated protected area, the use of endemic or native woods is prohibited, and the extraction and application of lava stone are heavily restricted and subject to rigorous oversight.

As a result, construction materials in the Galápagos are imported exclusively from mainland Ecuador, with concrete and cement blocks being the most prevalent due to their cost-effectiveness and ease of use. In the absence of alternative technologies and locally sourced materials, the architectural adaptation of the built environment to the archipelago's unique context can be understood primarily through the formal qualities of the structures themselves. This is assessed through a typological analysis of the buildings, which reveals patterns of spatial organization, material use, and morphological evolution over time.

Method of Analysis: Mapping the San Cristóbal Waterfront

Mapping the San Cristóbal waterfront enables a spatial and typological assessment of urban growth. This method allows for the quantitative analysis of the number, distribution, and architectural characteristics of buildings developed within this linear zone, offering insights into the morphological evolution of the island's-built environment.

The study focuses on the relationship between specific building typologies along the waterfront, with particular emphasis on the restaurant type. This urban strip, located at the edge of San Cristóbal Island, is primarily composed of tourism-oriented infrastructure that has emerged in response to increasing visitor flows.

Although the Galápagos Islands are globally recognized for their protected natural ecosystems, certain areas—particularly along the coast—have experienced ongoing urbanization. The built environment of the San Cristóbal waterfront reflects not only its proximity to the ocean but also the cumulative impacts of migration from diverse regions of mainland Ecuador.

Data for this study was collected along 17 blocks, encompassing a total of 1474 linear meters of the San Cristóbal waterfront (Fig. 4.2). Land use distribution varies by sector: the northern and southwestern ends are predominantly educational and residential in character, while the central section is largely oriented toward tourism.

Fig. 4.2 San Cristóbal Waterfront: map of documented blocks (Puerto Baquerizo Moreno, 2024) **Note. Map of the first 17 blocks along the San Cristóbal waterfront, identifying the main restaurant establishments from west to east: A. Nativo, B. La Góndola Restaurant, C. Giuseppe's, D. Sea Garden, E. Bar Restaurante Rosita, and F. Midori Sushi Pub. This map defines the primary zone of urban waterfront analysis for the study**

Fig. 4.3 Map of building uses on the ground floor, Waterfront of Puerto Baquerizo Moreno (2024) **Note. This map illustrates the distribution of ground-floor building uses along the San Cristóbal waterfront. Categories include residential, restaurants and coffee shops, souvenir and general tourism-related services, government infrastructure, and buildings under construction or used for educational purposes**

A total of 91 sites were surveyed and geolocated. Of these, 29 correspond to restaurants and coffee shops; 20 to tourism-related general services; 16 to tourism-oriented retail such as souvenir shops; 14 to educational institutions or buildings under construction; 10 to residential use; and 2 to government infrastructure.

Based on these findings, tourism-related typologies constitute the largest share of land use along the waterfront, representing 40% of all surveyed sites. Restaurant and café typologies follow at 31.1%, while educational or under-construction buildings account for 15.6%. Residential typologies comprise 11.1%, and government infrastructure represents the remaining 2.2% (Fig. 4.3).

Architectural Typologies: Categorizing and Understanding Patterns Along the Waterfront of San Cristóbal

The Galápagos Islands are widely regarded as a natural sanctuary, home to a unique array of plant and animal species. However, San Cristóbal lacks a vernacular architectural tradition, as there is no indigenous cultural group native to the archipelago. Instead, the island's population consists entirely of immigrants who arrived at different points in history, or of their descendants. This ongoing process of migration and settlement has led to a continuous process of adaptation, giving rise to diverse identities—referred to in this context as *sancristobaleños* (Ahassi 2007).

As discussed earlier, human settlement patterns in San Cristóbal have evolved over time, influenced by the presence of American naval forces, Ecuadorian migrants from various mainland cities, and the growing tourism industry. The built environment on the island reflects this layered history. Architectural typologies have adapted not only to the island's unique geographic and climatic conditions but also to the cultural backgrounds of its inhabitants.

Most of the construction materials used in San Cristóbal originate from mainland Ecuador, underscoring the island's material dependence. Architecture, in this case, becomes a reflection of the knowledge, practices, and aesthetics brought by these diverse populations (Ramos 2016). Along the waterfront, typological patterns are especially visible, shaped by both environmental constraints and the evolving cultural identity of the community. These patterns illustrate the complex relationship between cultural, social, and historical contexts and the emergence of specific architectural forms on the island.

For this analysis, 17 blocks along the waterfront of San Cristóbal were selected, and every building façade from Block 1 to Block 17 was documented. In total, 91 façades were redrawn. An example of this graphic documentation is shown in the image above (Fig. 4.4), which presents 43 façades from Blocks 2 to 10. In this diagram, all buildings identified as restaurants are highlighted in green.

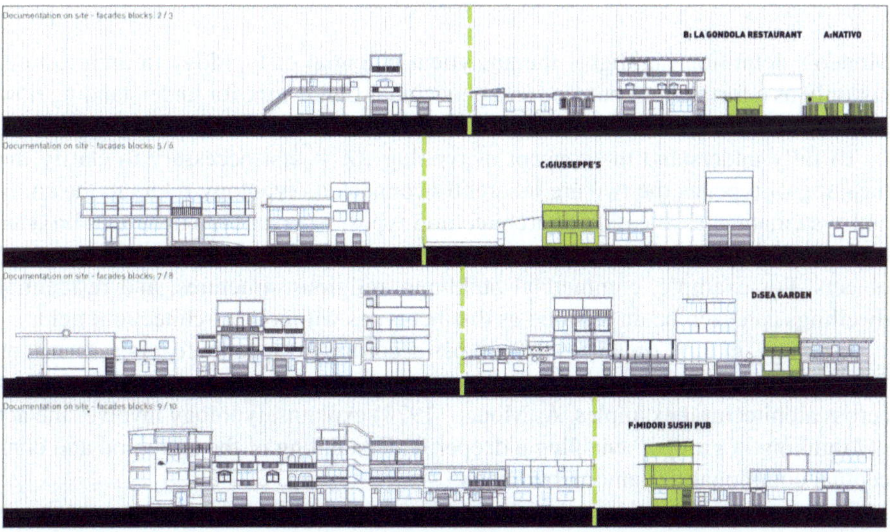

Fig. 4.4 Graphic example of on-site building facade documentation, Puerto Baquerizo Moreno Waterfront

Note. Photographic and drawn documentation of building facades along key waterfront blocks (Blocks 2–10) in Puerto Baquerizo Moreno. The image highlights architectural details and identifies prominent restaurant fronts, including A: Nativo, B: La Góndola Restaurant, C: Giuseppe's, D: Sea Garden, and F: Midori Sushi Pub

These restaurant typologies exhibit a consistent set of architectural features that reflect adaptation to the local territory, including elevated building bases, pilotis, and permeable front façades. These recurring characteristics suggest that architectural form is shaped in direct response to the environmental and spatial conditions unique to this coastal setting.

Consequently, the organic forms and transformation processes observed in these buildings are shaped by the specific geographic niche they inhabit. This supports the hypothesis that human settlements adapt to their environment through the construction process and the resulting built form.

The concepts of type and typology have long been central to architectural discourse. Throughout history, many renowned architects and theorists have explored these notions from a variety of perspectives. This relevance stems from a fundamental human tendency to classify and organize the world around us. In architecture, the impulse to categorize the built environment serves as a method to identify patterns, highlight distinctions, and reveal underlying structures within the constructed world.

Typology, in this context, is often regarded as a way to understand recurring spatial or formal characteristics. Several theorists have proposed that "type" can be seen as a foundational model or conceptual essence of architectural form. As Rafael Moneo (1978) explains:

> *What is type? It can most simply be defined as a concept which describes a group of objects characterized by the same formal structure. It is neither a spatial diagram nor the average of a serial list. It is fundamentally based on the possibility of grouping objects by certain inherent structural similarities. It might even be said that type means the act of thinking in groups.*

Moneo's definition highlights the analytical potential of typology in architecture, not only as a formal tool but also as a conceptual framework for understanding how buildings relate to one another within broader systems.

To fully understand the concept of typology, it is first necessary to clarify the meaning of type, as the two are inherently connected. Typology refers to the study and systematic classification of architectural types, whereas type denotes a broader category characterized by recognizable attributes shared by a group of architectural objects. For example, commercial buildings, religious structures, and residential dwellings can each be understood as distinct types within the architectural field.

Typology, in turn, organizes these types according to specific features, forms, or functions, offering a framework for identifying recurring patterns and variations across architectural examples. As Moneo (1978) explains, typology involves the act of "thinking in groups," enabling a deeper understanding of the structural and conceptual commonalities among buildings.

The concept of type can thus be seen as a population of buildings sharing a coherent set of attributes—attributes that stem from a collective understanding of function, purpose, and appropriateness of form. Like taxonomy in the biological sciences, architectural typology aims to identify, describe, and classify built forms based on shared traits. Alterations to these defining characteristics lead to transformations in the type itself.

As Kropf (2001) notes, these transformations are often triggered by shifts in environmental conditions or user needs. As perceptions of functionality, necessity, or beauty evolve, so too does the architectural type. For instance, in domestic architecture, modifications to specific design features driven by environmental factors reshape the conceptual understanding of what a house is and how it fulfills its purpose. These modifications, in turn, inspire further changes in the architectural form and how needs are perceived by the inhabitants of those buildings.

Recognizing variations in environmental conditions alters the needs that users seek to satisfy, prompting a shift in how they understand what makes a building functional and meaningful. This evolving perspective can lead to a redefinition of the architectural type itself (Kropf 2001). For example, in the case of the house, changes to certain design features—introduced to respond to environmental demands—reshape the idea of what a house is and how it should serve its occupants. These initial adjustments often lead to further refinements, influencing both architectural form and the way users perceive their spatial and functional needs.

However, buildings do not transform themselves—humans conceptualize and implement these changes in response to different contexts. In this sense, buildings serve as the material expression of human adaptation to the environment, while groups of buildings represent a collective response to the same conditions.

Building on these ideas, this research focuses on a recurring building type found along the waterfront of Puerto Baquerizo Moreno, spanning Blocks 1 through 17: the restaurant, which accounts for 31.1% of the area surveyed.

The Restaurant Typology

One of the earliest and most well-known restaurants in Puerto Baquerizo Moreno is *Bar Restaurant Rosita*. In 1945, María Celia Rumipamba migrated to the Galápagos Islands from Ambato, a city in Ecuador's Andean valley. As a new settler, she began preparing lunches for both residents and the established maritime community as a means of earning a livelihood. In doing so, she became one of the pioneers of San Cristóbal's food service industry, offering meals that evoked the comforting flavors of home-cooked Ecuadorian cuisine.

Three generations later, her granddaughter continues the family legacy. *Bar Restaurant Rosita* has since become a landmark, deeply embedded in the gastronomic and cultural history of Puerto Baquerizo Moreno.

This restaurant represents a significant point of departure for understanding the relationship between the early settlers of San Cristóbal—families who sought to generate income through small-scale entrepreneurship. This socio-economic context is one of the key reasons why the restaurant typology became one of the most prevalent architectural forms on the island. The arrival of the U.S. Navy in 1942 further accelerated the need for food services, as new naval personnel required supporting infrastructure. From that moment to the present day, the continuous influx of visitors—now primarily tourists—has sustained a high and growing demand for food establishments, reinforcing the restaurant as a dominant typology in the urban fabric of Puerto Baquerizo Moreno.

Along the waterfront, between Blocks 1 and 17, 31.1% of the buildings correspond to the restaurant typology. The graphic above (Fig. 4.5) helps to identify and analyze several of the defining characteristics common to this typology. The image features six case studies: A. Nativo, B. La Góndola Restaurant, C. Giuseppe's, D. Sea Garden, E. Bar Restaurante Rosita, and F. Midori Sushi Bar.

These three-dimensional illustrations offer insight into how these buildings interact with their immediate environment and highlight recurring architectural features. Among the shared elements observed are:

1. *The Building Base*: In coastal areas such as the waterfront of San Cristóbal, elevating buildings on a defined base is a widely adopted strategy to mitigate the risk of flooding throughout the year. This practice reflects an adaptation to the island's conditions. Typically, these building bases are constructed using locally available materials such as compacted earth, poured concrete, and volcanic stones sourced from the surrounding terrain.
2. *The Open Facade*: In San Cristóbal, precipitation is high between December and June and significantly lower from July to November. The hottest months occur between January and May, with temperatures reaching up to 28 °C, while cooler

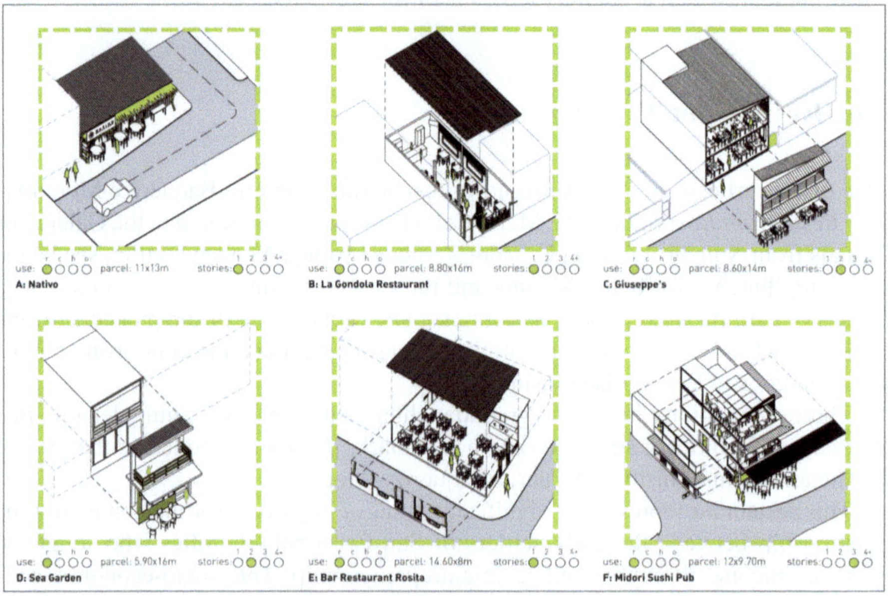

Fig. 4.5 Waterfront main restaurants documentation, Puerto Baquerizo Moreno (2024)
Note. Architectural documentation of six principal restaurant buildings along the San Cristóbal waterfront. The figure includes parcel dimensions, number of stories, and use classifications for: A. Nativo, B. La Góndola Restaurant, C. Giuseppe's, D. Sea Garden, E. Bar Restaurante Rosita, and F. Midori Sushi Pub

temperatures ranging from 20 °C to 24 °C are typical from June to December (Meteoblue 2024). Despite seasonal fluctuations, the climate remains predominantly humid, making open facades an effective architectural strategy. This approach reduces reliance on air conditioning or other electrical cooling systems, thereby lowering energy consumption and operational costs. Moreover, open facades enhance user experience by fostering a direct connection to natural elements such as ocean breezes and ambient coastal sounds. In multi-story buildings, this design also provides expansive views of the coastline and the sea.

3. *The Core Services*: The concentration of sanitary, electrical, drainage, and duct systems into a single vertical core follows a logic of functionality and efficiency. In most restaurants, this results in the kitchen and public restrooms being grouped in a specific, centralized area. This approach is both intuitive and cost-effective, as consolidating essential services in one location reduces construction costs and simplifies future maintenance. Additionally, vertical circulation elements—such as staircases—are often positioned adjacent to this core. This strategy frees up peripheral space within the building, allowing greater flexibility for dining areas and other programmatic functions.
4. *The Pilotis*: Within the tradition of modern architecture, pilotis have been widely used not only for structural support but also to enable flexible open-plan designs by freeing up the ground plane. In the case of San Cristóbal, however, the restaurants analyzed demonstrate that pilotis are employed primarily for structural necessity rather than for spatial liberation. The structural grid typically ranges from 3.5 to 5.5 m, accommodating the needs of each building's layout. The most used materials for these supports are metal and reinforced concrete, reflecting both local availability and the need for durability in the island's humid, coastal environment.

All these characteristics are inherently form-related, arising in direct response to the island's climatic conditions and the specific needs of users—on one hand, the restaurant owners, who prioritize cost-efficiency and spatial optimization, and on the other, the clients, who value a sensory connection to the sea and the broader island experience. Notably, these architectural features have emerged in San Cristóbal despite the absence of a deeply rooted indigenous architectural tradition and the scarcity of locally sourced construction materials. Together, these buildings exemplify a pragmatic and place-based adaptation to the unique environmental and cultural conditions of the Galápagos Archipelago.

Based on the documentation and analysis of 29 restaurants and coffee shops, six primary architectural typologies have been identified and are illustrated in Fig. 4.6. These typologies are categorized as follows:

- *Typology 1* (9 restaurants): Single-story buildings characterized by an open façade and the use of pilotis.
- *Typology 2* (1 restaurant): The least common typology, distinguished by a building surrounded by dense vegetation and a frontal garden.
- *Typology 3* (2 restaurants): Two-story buildings with open ground floors and open upper-floor façades.

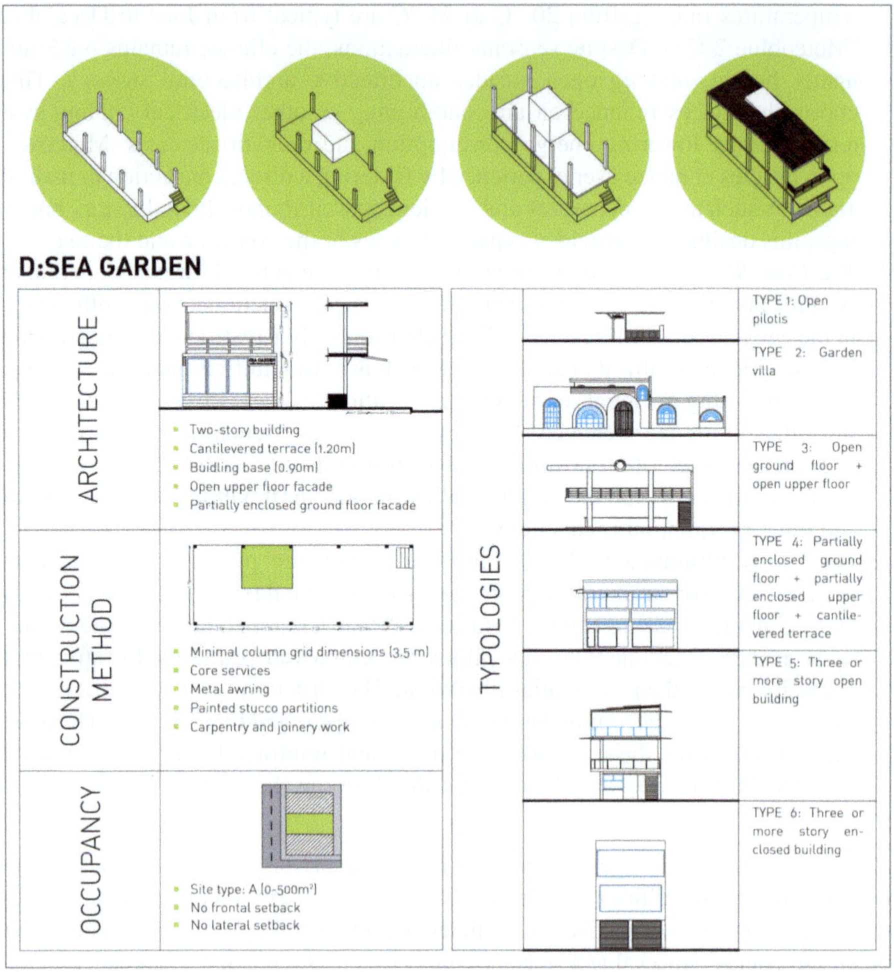

Fig. 4.6 Illustration of a representative restaurant typology on the Waterfront, Puerto Baquerizo Moreno (2024)
Note. Analytical drawing of a typical waterfront restaurant building, exemplified by Sea Garden. The illustration presents architectural elements such as a cantilevered terrace, building base, open and partially enclosed facades, column grid dimensions, and construction materials. The typology reflects spatial and material characteristics common among waterfront restaurants, with no frontal or lateral setbacks and an open upper-floor layout

- *Typology 4* (11 restaurants): Similar to Typology 3, but with a partially enclosed ground floor and a fully enclosed upper floor featuring a cantilevered terrace.
- *Typology 5* (2 restaurants): Multi-story buildings with three or more levels, each featuring open façades.
- *Typology 6* (4 restaurants): Enclosed buildings with three or more stories, lacking the open façade characteristics of the other types.

Typologies 1 and 4 are the most prevalent among the documented cases in San Cristóbal, representing most restaurant establishments. These typologies are defined by several recurring architectural strategies: a raised building base of approximately 90 centimeters, a partially enclosed ground-floor façade, a structural grid of around 3.5 m, and a centralized concrete core that houses essential services such as bathrooms and kitchens. Additionally, most of these restaurants are built without frontal or lateral setbacks, maximizing the use of available plot space. Together, these defining features contribute to the distinct and locally adapted character of San Cristóbal's restaurant architecture.

Conclusion

This chapter has examined the evolution of San Cristóbal's urbanized waterfront by analyzing its urban morphology and documenting one of its most predominant architectural typologies: the restaurant. Through this lens, a representative cross-section of the architecture of Puerto Baquerizo Moreno has been revealed, illustrating how environmental conditions, cultural dynamics, and historical developments have shaped the built environment of this coastal settlement.

These results underscore the concept of architectural endemism, echoing Darwin's theory of evolution. Just as species adapt to their environments, the restaurant typologies in San Cristóbal have evolved in response to local constraints and opportunities. In the specific case analyzed, this adaptation is expressed by a raised building base, a centralized concrete service core, open façades, structural pilotis, and full lot occupancy without frontal or lateral setbacks. This typological evolution reflects how human ingenuity can align with environmental conditions, offering important insights for the development of sustainable urban forms in ecologically sensitive regions.

This research opens the door to investigating typological adaptations within specific geographical contexts, as exemplified by the Galápagos. Beginning with San Cristóbal, future studies could explore additional building types within the archipelago using this framework to understand the emergence of context-specific architectural forms. For instance, the urban hotel typology on Santa Cruz Island has undergone a significant evolution since the 1950s, beginning with the establishment of "Hotel Galápagos." This hotel displays recurring architectural features that can be identified and analyzed as typological patterns. A comparative study of similar building types in other regions with comparable characteristics could offer valuable analytical insights, helping to envision how the built environment in the Galápagos Islands might evolve in ways that interact more appropriately and sensitively with nature.

Just as organic forms in the Galápagos did not evolve from a single cellular organism within the islands but instead arrived and adapted to the geographical context due to its unique characteristics (Valle 2010), architectural forms have similarly evolved. These forms arrived at different points in time and were transformed to better respond to the specific environment of the Galápagos as a form of human adaptation. Typological comprehension goes beyond a simple classification of built forms; it investigates how people have responded to their surrounding environment through the things they have built and found useful. Within this context, this research contributes to the ongoing debate on how buildings should be designed to harmonize with the unique environment of the Galápagos, fostering a symbiotic relationship between human settlements, culture, history, and architecture.

References

Ahassi H (2007) Lo galapagueño, los galapagueños. Revista de Antropología Experimental 7:169–176

Alexander C (1964) Notes on the synthesis of form. Harvard University Press

Darwin C (1859) On the origin of species by means of natural selection, or the preservation of favoured races in the struggle for life, 1st edn, John Murray

Granda L, Salazar M, Salazar GC (2013) Population and migration in Galápagos. In DPNG, CGREG, FCD, & GC (eds), Informe Galápagos 2011–2012. Dirección del Parque Nacional Galápagos, pp 44–51

Huxley J (1944) On living in a revolution. Chatto & Windus

Idrovo H (2005) Galápagos: Huellas en el paraíso. Librimundi

Instituto Nacional Galápagos (2005) Plan de manejo del Parque Nacional Galápagos. Presidencia de la República del Ecuador

Kvan T, Karakiewicz J (eds) (2018) Urban Galápagos: transition to sustainability in complex adaptive systems. Springer. https://doi.org/10.1007/978-3-319-99534-2

López JE (2024, October 2) Zonificación tipológica en las Islas Galápagos: El caso de la Isla de Santa Cruz. In VIII Congreso Internacional ISUF-H Valencia 2024 – Formas urbanas diversas para espacios en recomposición. https://doi.org/10.4995/ISUFh2024.2024.17930

Losos JB, Arnold SJ, Bejerano G, Brodie ED III, Hibbett D, Hoekstra HE et al (2013) Evolutionary biology for the 21st century. PLoS Biol 11(1):e1001466. https://doi.org/10.1371/journal.pbio.1001466

Meteoblue (2024) Meteorological variables. https://www.meteoblue.com/en/weather/historyclimate/climatemodelled/san-crist%C3%B3bal-island_ecuador_3651946

Moneo R (1978) On typology. Oppositions 13:23–30

Ramos D (2016) Género a-islado: Una re-lectura del territorio desde las prácticas cotidianas. Congreso Arcadia 4:159–169

Rudofsky B (1965) Architecture without architects: an introduction to non-pedigreed architecture. Museum of Modern Art

Sánchez P (2022) Indigenous settlers of the Galapagos: conservation law, race, and society. Lexington Books

Stahl PW, Astudillo FJ, Jamieson RW, Quiroga D, Delgado F (2020) Historical ecology and archaeology in the Galápagos Islands. University Press of Florida

Valle C (2010) Science and conservation in the Galápagos Islands. In: Walsh SJ, Mena C (eds) Science and conservation in the Galápagos Islands: frameworks and perspectives. Springer, pp 1–22

Chapter 5
Freshwater in the Galápagos Islands: A Multifaceted Battle from Water Scarcity to Wastewater Management

Aracely Zambrano-Romero, Andrea C. Encalada, Daniel Escobar-Camacho, Daniela Rosero-López, Diana Pazmiño, Doménica Arguello, Henry Vásconez, María José Ayala, Pedro Gallegos, Ruth Fernanda Morales, Susana Hernández, Kory Russel, and Valeria Ochoa-Herrera

Abstract The Galápagos Islands face complex challenges at the intersection of freshwater ecosystems, sustainable water resources use, and public health. The growing population of residents and tourists strains freshwater resources, prompting a critical reassessment of water management and the adverse impacts on the environment. This chapter offers a comprehensive review of scientific literature to assess the current state and future prospects of freshwater ecosystems and wastewater

A. Zambrano-Romero · M. J. Ayala
Colegio de Ciencias e Ingenierías, AQUA-BIO Lab, Universidad San Francisco de Quito, Quito, Ecuador

A. C. Encalada
Laboratorio de Ecología Acuática, Global Research and Solutions Center, Universidad San Francisco de Quito, Quito, Ecuador

Galápagos Science Center, Universidad San Francisco de Quito and University of North Carolina at Chapel Hill, San Cristóbal, Ecuador

D. Escobar-Camacho · D. Rosero-López
Laboratorio de Ecología Acuática, Global Research and Solutions Center, Universidad San Francisco de Quito, Quito, Ecuador

D. Pazmiño
Galápagos Science Center, Universidad San Francisco de Quito and University of North Carolina at Chapel Hill, San Cristóbal, Ecuador

D. Arguello · S. Hernández
Instituto de Microbiología, Universidad San Francisco de Quito USFQ, Quito, Ecuador

H. Vásconez · P. Gallegos · R. F. Morales
Colegio de Ciencias de la Salud, Universidad San Francisco de Quito, Quito, Ecuador

K. Russel
College of Design and Environmental Studies Program, University of Oregon, Eugene, OR, USA

J. Dunn Insua et al. (eds.), *Human Galapagos*, Social and Ecological Interactions in the Galapagos Islands, https://doi.org/10.1007/978-3-032-08800-0_5

management in the Galápagos Islands. Beyond summarizing existing evidence, it critically analyzes the social-ecological implications of prevailing water management strategies, addressing key knowledge gaps, ecological and institutional constraints, and the urgent need for actionable, science-based solutions. The chapter examines the drivers of freshwater vulnerability and evaluates the diverse responses—both positive and negative—that have shaped water governance in the Galápagos. A central theme is the concept of limits, emphasizing that freshwater resources in the Islands are not only scarce but finite, underscoring the need for careful management in a delicate, ecologically unique environment. Freshwater ecosystems are integral to the broader socio-ecological system, supporting endemic biodiversity, enhancing island resilience, and possessing inherent conservation value. This holistic view stresses their importance beyond merely serving human needs. Finally, the chapter underscores that addressing water-related challenges in the Galápagos Islands, from water scarcity to wastewater management, requires a multifaceted and multidisciplinary approach. This approach must integrate environmental science, biology, hydrology, engineering, and public health, guided by technical feasibility, scientific rigor, and ethical considerations, with a strong emphasis on environmental justice to ensure sustainable water governance for present and future generations in the Galápagos Islands.

Keywords Freshwater ecosystems · Wastewater · Human health · Biodiversity · Conservation · Galápagos

Background and Objectives

The Galápagos Archipelago, located approximately 900 kilometers off the coast of mainland Ecuador in the eastern Pacific Ocean, is one of the world's most iconic natural laboratories. Renowned for its extraordinary biodiversity, high levels of endemism, and striking geological features, the archipelago was designated a UNESCO World Heritage Site for both its terrestrial and marine ecosystems. As one of the youngest volcanic island chains on Earth, the Galápagos continue to shape our understanding of island biogeography, evolution, and ecosystem dynamics (Mena et al. 2020). The islands gained global scientific prominence through Charles Darwin's groundbreaking observations, which contributed to the formulation of the theory of evolution by natural selection—arguably one of the most transformative ideas in modern science (Cordovez et al. 2024).

V. Ochoa-Herrera (✉)
Colegio de Ciencias e Ingenierías, AQUA-BIO Lab, Universidad San Francisco de Quito, Quito, Ecuador

Galápagos Science Center, Universidad San Francisco de Quito and University of North Carolina at Chapel Hill, San Cristóbal, Ecuador
e-mail: vochoa@usfq.edu.ec

Beyond their symbolic and scientific value, the Galápagos provide critical ecosystem services to a growing local population. Human settlements are concentrated on just four of the thirteen major islands—Santa Cruz, San Cristóbal, Isabela, and Floreana—all of which are the only islands that possess natural freshwater ecosystems. In contrast, the remaining islands lack permanent freshwater sources, underscoring the extreme hydrological constraints of the archipelago and the central role freshwater availability plays in shaping both human and ecological sustainability (Burbano and Meredith 2021; Vásquez et al. 2022; Alava et al. 2023) (Fig. 5.1).

In 2022, the Galápagos Islands were home to over 28,000 residents across three main cantons: Santa Cruz (17,233 inhabitants, including Floreana), San Cristóbal (8300), and Isabela (3050) (INEC 2022). The Islands also host a fast-growing floating (tourism) population driven mainly by interest in nature and wildlife (Mena et al. 2020; Alava et al. 2023). A total of 267,688 tourists visited Galápagos in 2022, nearly surpassing the number of visitors in 2019 before the COVID-19 pandemic, showing an impressive recovery for the tourism industry (Burbano and Meredith 2021).

Freshwater scarcity is one of the most critical and persistent challenges in the Galápagos Islands. The availability of freshwater is inherently constrained by the islands' young volcanic origin, fractured basaltic geology, and a climate marked by seasonal rainfall and high evaporation (Auken et al. 2009). Only four inhabited islands host natural freshwater ecosystems, and even there, sources are limited,

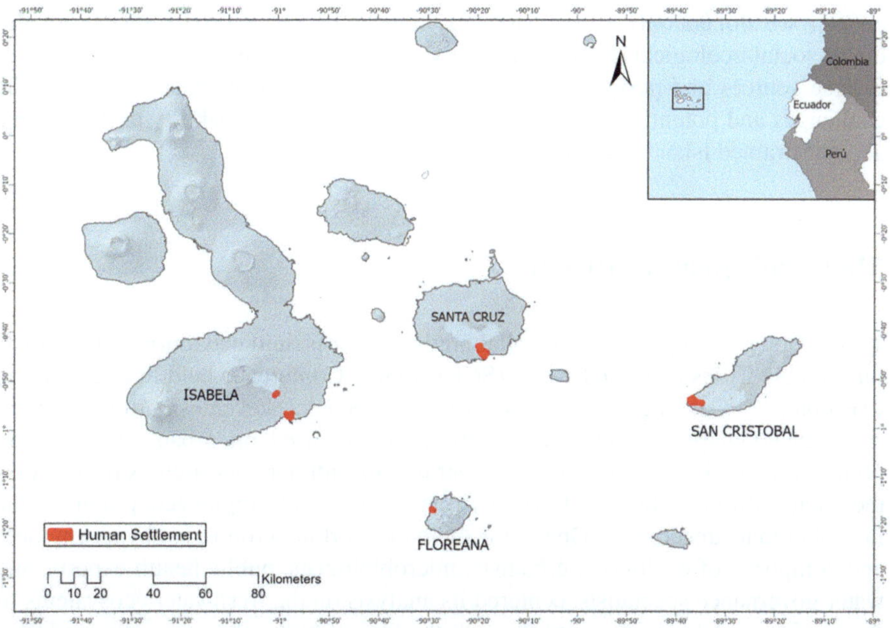

Fig. 5.1 Map of human settlement in the Galápagos Islands: San Cristóbal, Santa Cruz, Isabela and Floreana

often brackish, and vulnerable to contamination. Despite varied water supply strategies—including rainwater harvesting, brackish groundwater extraction, and desalination—water quality and quantity remain precarious (Mateus et al. 2019; Riascos-Flores et al. 2021; Grube et al. 2020; Badhwa et al. 2022; Pryet et al. 2012; Schmitt et al. 2018). Inadequate infrastructure and poorly maintained systems expose local populations to health risks and compromise freshwater-dependent ecosystems (Fernandes Reyes 2022; Grube et al. 2020, Mateus et al. 2020). Moreover, wastewater mismanagement and the absence of robust treatment systems further degrade environmental quality, both inland and along the coasts (Vásquez et al. 2022; Riascos-Flores et al. 2025). Governance approaches tend to be centralized and technocratic, often overlooking the lived realities and traditional knowledge of local communities. Public participation in water planning is limited, resulting in policies that struggle to address real needs (Mateus et al. 2020; Vásquez et al. 2022; Riascos-Flores et al. 2025). Critically, the biophysical limits of the islands—small size, fragile ecosystems, finite water availability—are frequently ignored in development agendas, raising serious concerns about long-term sustainability and the unchecked expansion of human settlements and tourism.

Against this backdrop, this chapter offers a comprehensive review of the scientific literature to explore the current state and future prospects of freshwater ecosystems and wastewater management in the Galápagos Islands. The following sections synthesize insights from interdisciplinary research—spanning ecology, engineering, hydrology, public health, and policy—to unpack the complex dynamics of water scarcity, infrastructure, governance, and health-related risks. Throughout the chapter, we aim not only to summarize existing evidence but also to reflect critically on the social-ecological implications of prevailing approaches. By weaving together diverse sources and perspectives, we contribute to a deeper understanding of the challenges and potential solutions for achieving water sustainability in this unique and constrained island system.

Methodological Approach

This chapter emerged from a collaborative process initiated during the APRU Sustainable Cities and Landscapes (SCL) Annual Conference held in 2023 on San Cristóbal, Galápagos (https://www.usfq.edu.ec/es/eventos/apru-scl-annual-conference-Galápagos-2023). The conference—focused on urban challenges in protected natural areas—brought together interdisciplinary researchers from across the Pacific Rim to reflect on how cities can coexist with fragile ecosystems in the face of climate uncertainty. One of the thematic working groups, led by an engineer and composed of ecologists, urbanists, microbiologists, public health experts, and water governance specialists, centered its analysis on the freshwater ecosystems of the Galápagos Islands—viewed through the lens of both urban pressures and ecological conservation.

Following the initial workshop, the group undertook a comprehensive literature review to assess the state of knowledge on freshwater availability, water quality, wastewater treatment, governance, and socio-environmental challenges across the inhabited islands of the Archipelago. Bibliographic searches were conducted using Web of Science and Scopus databases, focusing on publications in English and Spanish and using terms such as "water," "wastewater," "health," "biodiversity," and "Galápagos." The review included peer-reviewed articles, book chapters, technical reports, conference proceedings, and relevant grey literature. A curated set of approximately 60 publications formed the basis for thematic analysis, following a transparent and systematic framework (Gough et al. 2017).

However, this chapter is more than a synthesis of existing literature. It represents a collective and critical reflection on the major knowledge gaps, the ecological and institutional limitations, and the urgent need for actionable, science-informed solutions. The group examined both the drivers of freshwater vulnerability and the responses—positive and negative—that have shaped water governance in the Galápagos Islands. Special emphasis was placed on the concept of limits: the recognition that freshwater in Galápagos is not only scarce but finite, and that water management must occur within the boundaries of a highly sensitive and ecologically unique island system.

This approach acknowledges that freshwater ecosystems in the Galápagos are not merely service providers for human use, but vital components of a complex socio-ecological system that supports endemic biodiversity, underpins island resilience, and carries intrinsic conservation value. Thus, the reflections and proposals offered throughout the chapter are guided not only by technical feasibility and scientific rigor, but also by ethical considerations and the imperative to ensure environmental justice for current and future generations of Galapagueños.

Freshwater Ecosystems of Galápagos

Freshwater ecosystem studies on lentic and lotic environments have focused on resources for water provision. However, few investigations have addressed freshwater ecosystem dynamics with water accessibility (Karnauskas et al. 2014; Percy et al. 2016). In the Galápagos Islands, freshwater is available in lakes, temporary and permanent ponds, and permanent springs present only in geologically older islands (Fig. 5.2) (Violette et al. 2014; Schmitt et al. 2018). This is the case of the geologically oldest island, San Cristóbal (max. elevation 712 m), that has a unique aquatic ecosystem across the entire Archipelago, considered a highland lake on top of a confined aquifer. Other inhabited islands such as Floreana, (max. elevation 567 m) and Santa Cruz (max. elevation 845) exhibit ponds and temporary streams that form in ravines in the rainy season (from December to May). However, due to high permeability in the soils of these islands, streams do not reach the sea (Paltán et al. 2021). San Cristóbal is the only island with dyke and perched aquifers, while the other islands exhibit only perched varieties (Pryet et al. 2012; Violette et al.

Fig. 5.2 Freshwater and brackish water ecosystems in the Galápagos Islands. Legend: (**a**) Lake (freshwater) of El Junco (at 700 msn) in San Cristóbal, (**b**) Pond (freshwater) in "Ranco Los Manzanillos", a touristic venue in Santa Cruz, (**c**) Las Diablas lagoon (brackish water), an interconnected systems of lagoons in Isabela, (**d**) Lake (brackish water) in Beagle Crater in Isabela. (Photos: Courtesy of Karla Barragán)

2014). The age of each island, geomorphology, and near-surface lithology all have significant influence on the hydrological characteristics, as well as the distribution and occurrence of freshwater ecosystems (Violette et al. 2014).

Currently, sources for freshwater include groundwater, desalination and fog or rainwater harvesting, with rainfall being the key supplier of most sources (Echeverría et al. 2020; Badhwa et al. 2022; Vásquez et al. 2022; Thompson et al. 2023). During the wet season, rainfall is more frequent and intense, while the dry season, from June to November, is characterized by reduced precipitation levels (Paltán et al. 2021). Rainfall provides essential moisture for terrestrial ecosystems, replenishes temporary surface water bodies, and recharges groundwater aquifers in San Cristóbal. However, rainfall in the Galápagos is highly variable, influenced by seasonal weather patterns such as El Niño and La Niña events (Escobar-Camacho et al. 2021; Paltán et al. 2021; Mateus and Quiroga 2022; Thompson et al. 2023).

The Galápagos Islands support a network of predominantly seasonal surface water bodies, including rivers, streams, ponds, and lakes. While surface water is relatively scarce compared to other regions, these water bodies play a crucial role in supporting freshwater-dependent species and providing habitat for aquatic life (Mena et al. 2020; Dueñas et al. 2021). Between 37% and 64% of the annual rainfall is lost to evapotranspiration, with higher values observed in San Cristóbal than other islands (Escobar-Camacho et al. 2021). Therefore, most streams in the Galápagos are ephemeral, flowing only during the wet season and drying up during the dry season (Alomía Herrera et al. 2022; Paltán et al. 2023). Only a few streams in San Cristóbal are permanent (i.e., Encañada Milton Aguas, Encañada Jocotoco, Cascada Jatun Sacha) (Barragán et al. 2025; Cárdenas López 2019). In Floreana (e.g., Wittemeyer) and Santa Cruz (e.g. Bellavista) permanent spring exist and are used entirely to supply freshwater to the population (Barragán et al. 2025; Cárdenas López 2019).

Moreover, surface runoff from higher inland territories to the coast contributes nearly 10% of the annual water flows, while interflow adds merely 6%. Notably, in San Cristóbal, surface runoff accounts for a mere 1% of the total streamflow, despite this island having perennial streams (Escobar-Camacho et al. 2021; Paltán et al. 2021). Ponds and lakes are also present, serving as essential freshwater habitats and breeding grounds for endemic species (Escobar-Camacho et al. 2021; Dueñas et al. 2021; Thompson et al. 2023).

Over the years, significant progress has been made in understanding the diversity of freshwater and brackish ecosystems in the Galápagos Islands. Early studies focused on paleolimnology (Bush et al. 2010), while recent research has uncovered new algae species (Bak et al. 2017), documented human-associated insect spread (Causton et al. 2006), and highlighted the role of aquatic habitats in supporting diverse fauna (Gerecke et al. 1995). These ecosystems are shaped by volcanic origins, geological age, elevation, and hydroclimate, yet remain underrepresented in conservation strategies despite their importance for biodiversity, tourism, and ecosystem services. A recent guide has identified 138 aquatic invertebrate species (including insects and decapods—found in freshwater and brackish ecosystems across the Archipelago) of which 28.3% (39 species) are endemic, 33.3% are native (46 species), and 23.9% (33 species) are introduced (Barragán et al. 2025), raising ecological concerns and highlighting the need for monitoring and prevention strategies. The order Diptera (flies), with 11 families and approximately 90 species, is the most diverse group associated with freshwater and brackish environments in the Galápagos. Following this, Hemiptera (true bugs) includes seven families and 18 species, Coleoptera (beetles) comprises six families and 15 species, and Odonata (dragonflies) is represented by three families and eight species. For order Decapoda (crustaceans), three families of freshwater and brackish shrimp and crabs have been documented. While some marine crab species may occasionally inhabit brackish environments, limited ecological data restricts the species list to the three most thoroughly studied families. These findings emphasize the need for continued research -not only to build robust baselines for conservation and monitor invasive species, but also to better understand the ecological functions and services freshwater and brackish ecosystems provide. As freshwater habitats in the Galápagos remain among the least studied components of the archipelago, expanding knowledge in this area is essential for safeguarding both biodiversity and the health of the Island ecosystems.

Each island has different types of fresh- and brackish-water sources. The Lake El Junco watershed is the main water source for San Cristóbal's population (Echelpoel et al. 2019; Grube et al. 2020; Echeverría et al. 2020). The current supply does not provide consistent and pressurized distribution to households (Echelpoel et al. 2019; Grube et al. 2020). As a result, the system is prone to contamination from intermittent flow and instances of re-growth during distribution and household storage. On Floreana, the primary water source is a natural pond replenished during the rainy season (Rhea et al. 2023).

Groundwater represents a vital source of freshwater in the Galápagos Islands, particularly in areas where surface water is limited or unreliable such as in Santa Cruz and Isabela. The municipal desalination plants installed in these two islands draw water from brackish groundwater and remove salts and impurities to produce potable water. In Santa Cruz, assessments of coastal and groundwater sources have revealed their unsuitability for drinking and domestic use at certain inland locations (Fernandes Reyes et al. 2017; Mateus et al. 2019; Re et al. 2023). In Isabela, water quality is impacted by the infiltration of seawater (Riascos-Flores et al. 2021; Badhwa et al. 2022).

Desalination plays a critical role in providing freshwater for drinking, agriculture, and other human activities in areas where traditional water sources are insufficient (Vásquez et al. 2022). However, desalination can be energy-intensive, expensive, and produces high salinity brine as a waste product, posing challenges in terms of sustainability and cost-effectiveness (Mateus et al. 2020; Vásquez et al. 2022; Ayala-Pico et al. 2023). In terms of quantity, according to Paltán and collaborators, approximately 34% of the available water in the Galápagos Islands seeps into the ground (Paltán et al. 2021).

In some parts of the Galápagos Islands such as in Santa Cruz and San Cristóbal Islands, fog and rainwater harvesting are practiced as sustainable methods for collecting and storing freshwater (Pryet et al. 2012; Schmitt et al. 2018; Echeverría et al. 2020). Rainwater is collected from rooftops and other impervious surfaces and stored in cisterns or tanks for later use. Rainwater harvesting reduces reliance on external water sources and can supplement household water supplies, particularly during the wet season (Echeverría et al. 2020). In Santa Cruz Island, the rainwater harvesting capacity has been calculated at 4000 m^3 across 2000 households. This capacity can meet 72% of the total water demand in Puerto Ayora, assuming slow population growth scenarios (Reyes et al. 2017). The availability of water obtained through rainwater and fog harvesting depends heavily on seasonal rainfall patterns and the capacity of storage infrastructure (Schmitt et al. 2018).

Finally, freshwater resources on the Galápagos Islands are both finite and scarce; therefore, careful management is essential within this fragile and ecologically unique environment. These freshwater ecosystems are vital components of the broader socio-ecological system, supporting endemic biodiversity, strengthening island resilience, and possessing inherent conservation value. This holistic view highlights their importance beyond simply meeting human needs, calling for a balanced and sustainable approach to their protection and stewardship.

Water and Human Health

In recent years, there has been growing recognition of the need to understand the interconnectedness of human, animal, and environmental health, aligning with the One Health approach. This integrated framework aims to promote the sustainable use of resources while optimizing the well-being of people, animals, and ecosystems (Nicholas et al. 2020; Thompson et al. 2023).

The availability of freshwater sources influences public health, serving as essential reservoirs for drinking water, agricultural needs, and recreational activities. In regions with water scarcity, challenges in maintaining hygiene contribute to malnutrition, particularly among children (He et al. 2021). Islands and archipelagos face numerous challenges in freshwater access, relying on limited local sources prone to shortages and contamination (Violette et al. 2014; Percy et al. 2016). Geographic isolation complicates healthcare response and climate change exacerbates scarcity (Heads and Grehan 2021). Sustainable water management and community resilience are vital for safeguarding health in these settings, emphasizing the need for tailored approaches to water management (Wingfield et al. 2021; Fernandes Reyes 2022; Mateus and Quiroga 2022; Re et al. 2023). However, community and local actors are not always involved in the decision-making process or in the enforcement measures to address water quality degradation (Wingfield et al. 2021). In the Galápagos, inhabitants are not confident in the quality of tap water and prefer to consume bottled water instead (Mateus et al. 2019; Vásquez et al. 2021, 2022; Fernandes Reyes 2022).

The Galápagos Islands face intricate challenges at the intersection of drinking water sources and public health (Nicholas et al. 2020; Grube et al. 2020; Thompson et al. 2023). The explosive pre-2000s population growth led to spontaneous human settlements, causing rapid modifications to the water supply and an escalating need for public services. The demand for resources such as water, energy and food coupled with minimal environmental awareness, has had a negative impact on the delicate balance of these ecosystems (Nicholas et al. 2020). This highlights the critical importance of adopting an integrated approach to the water-energy-food nexus (WEF nexus), ensuring that the interconnected needs of these vital resources are met sustainably while preserving the unique biodiversity of the Archipelago where freshwater is scarce and finite.

Various sources, including sewage, agriculture, livestock activities, and waste accumulation, contribute to contamination (Nicholas et al. 2020; Thompson et al. 2023). The Islands' geographical constraints exacerbate these issues, as limited freshwater resources heighten dependence on local sources, affecting not only Santa Cruz but also other inhabited islands like San Cristóbal and Isabela (Grube et al. 2020; Badhwa et al. 2022; Fernandes Reyes 2022). The booming tourism industry, while economically beneficial, increases the pressure on already collapsed local water resources, as hotels and other services drive higher demand. Likewise, the increased need for food strains an agricultural system that has already stretched beyond capacity due to limited arable land and irrigation water. Improper farming practices, such as pesticide overuse contaminate freshwater, highlight the need for more sustainable methods of safeguarding water quality (Riascos-Flores et al. 2021). Studies conducted on San Cristóbal Island report that public health is directly affected by limited access to clean water. Poor water quality is linked to a higher incidence of diarrheal diseases and other illnesses transmitted through food and water (Houck et al. 2020; Thompson et al. 2020; Nicholas et al. 2020). Further, tourists may experience these diseases, primarily attributed to the consumption of contaminated food and water, while cases of vector-borne diseases like dengue fever or Zika virus have become prevalent. Recent disease control efforts have concentrated on combating vectors, overlooking the issue of stagnant water sources, which serve as breeding grounds for these disease carriers (Mosquera et al. 2023). Efforts to safeguard freshwater ecosystems are not only essential to protect biodiversity but also play a vital role in preserving public health, ensuring a sustainable future for both the environment and the inhabitants of these fragile Archipelago (Nicholas et al. 2020). In addition to public health concerns, poorly treated wastewater also poses risks to both terrestrial and aquatic biodiversity in the Galápagos, exposing endemic fauna to emerging contaminants, pathogens, and habitat degradation. Finally, one of the main motivations to understanding the intricate paths between water sources and fate of wastewater in aquatic ecosystems was the reported contamination of popular recreational spots which were flagged as unsafe for swimming due to poor wastewater disposal practices. Moreover, despite their compromised environmental status, certain areas remain open to tourism, potentially perpetuating a cycle of health risks (Nicholas et al. 2020; Thompson et al. 2023).

Wastewater Challenges in the Galápagos

In the Galápagos Archipelago, wastewater originates from diverse sources such as tourism, agriculture, domestic activities, maritime traffic, and rainwater run-off. Wastewater management practices differ significantly across the Galápagos Islands, and unfortunately, only limited initiatives have been implemented to address these pressing challenges. Efforts to manage wastewater have primarily focused on the use of septic tanks, the installation of wastewater treatment plants and the development of constructed wetlands. However, most wastewater remains untreated, posing a severe threat to both the ecosystem and the health of the local human population in this fragile environment.

Table 5.1 summarizes the information related to population and current wastewater treatment systems per island. Santa Cruz only has domestic septic tanks that connect individual houses, the city's sewage coverage level is only 3.3% and just 10% of those septic tanks function properly. The remaining septic tanks suffer from materials defects of poor maintenance. As a result, untreated or inadequately treated sewage water infiltrates the ground and contaminates the aquifers (Ragazzi et al. 2016; Mateus et al. 2019). The brackish water sourced from underground fissures exhibits high chloride concentrations (800 to 1200 mg/L) indicative of seawater

Table 5.1 Island, population and current wastewater treatment systems for Santa Cruz, San Cristóbal and Isabela

Island	Population[a]	Main Urban Areas	Current wastewater treatment systems	Observations
Santa Cruz	17,233	1. Puerto Ayora. 2. Bellavista. 3. Santa Rosa.	None[b]	No sewerage network[b,c] 97.2% connected to septic tanks–1.9% to the public sewerage network[b,c]
San Cristóbal	8300	1. Puerto Baquerizo Moreno. 2. El Progreso.	Conventional treatment facility operating since 2012, treating 6–8 L/s[b,c]	Existing system connected to 99% of households[b] Operational issues[b,c]
Isabela	3050	1. Puerto Villamil. 2. Tomas de Berlanga.	Small artificial wetlands with native aquatic plants operating since 2015[d]	Supported by Galápagos conservation Trust[d]

Modified from Mateus et al. (2020)
[a]INEC (2022)
[b]Mateus et al. (2019, 2020)
[c]Fernandes Reyes et al. (2017), Grube et al. (2020), Riascos-Flores et al. (2021), and Fernandes Reyes (2022)
[d]Walsh et al. (2010) and Badhwa et al. (2022)

Fig. 5.3 Wastewater management in San Cristóbal Island. Legend: (**a**) Central pumping station, (**b**) Wastewater treatment plant. (Photos: Courtesy of Valeria Ochoa-Herrera)

intrusion and the presence of *E. coli* (Fernandes Reyes et al. 2017; Mateus et al. 2019; Nicholas et al. 2020). San Cristóbal has a conventional treatment facility that began operating in 2012, treating 6–8 liters of wastewater per second (Fig. 5.3).

However, the plant's operation may have some longer-term sustainability issues and currently treats only a part of the domestic effluents generated on the Island (Fernandes Reyes et al. 2017; Mateus et al. 2019, 2020; Grube et al. 2020; Riascos-Flores et al. 2021; Fernandes Reyes 2022). Isabela has a small sewage treatment system in the form of artificial wetlands with native aquatic plants (Walsh et al. 2010; Badhwa et al. 2022).

Inadequate treatment and disposal of wastewater threaten the islands' biodiversity and public health (Tilmans et al. 2014). Contamination of water sources due to untreated wastewater introduces harmful substances, pathogens, and pollutants such as emerging contaminants. A study evaluating urban and associated natural water systems on Floreana, Isabela and Santa Cruz, revealed significant impacts, particularly on Santa Cruz, where approximately 39% of monitoring sites were classified as polluted according to the Basic Prati Indexm, with *E. coli* detected in 57% of the sites, and sewage contributing up to 57% of the NO_3^- in the receiving water bodies (Riascos-Flores et al. 2025). On Isabela, treated wastewater discharge contained high ammonium concentrations, *E. coli*, and critically low oxygen levels. On Floreana, the aquifer closest to the urban area showed nitrate contamination with up to 94% of NO_3^- originating from sewage, compared to just 11% on the other islands. Wastewater discharges not only disrupt the aquatic ecosystems but also alter downstream nutrient inputs to the marine environments. In fact, this nutrient imbalance, particularly the excessive accumulation of nitrogen and phosphorus, has triggered algal blooms and the formation of dead zones in marine areas (Mateus et al. 2019). These blooms deplete oxygen levels, suffocating aquatic life in the marine ecosystem and posing a significant threat to the thriving biodiversity in these waters. Beyond the marine environment, the repercussions extend to terrestrial ecosystems, affecting plants and land animals. The consequences of poor wastewater management include contaminated water seeping into the soil, impacting vegetation, disrupting food chains, and ultimately threatening the survival of species reliant on these ecosystems (Miller et al. 2022). Moreover, the repercussions of poor wastewater management persist throughout the Archipelago and are exemplified by popular recreational spots deemed unsafe for swimming (Ragazzi et al. 2016).

The cumulative contamination impacts and future climate change scenarios have significant implications for wastewater management strategies in the Galápagos Islands. Climate change is already impacting current wastewater infrastructure. Altered precipitation patterns can affect the volume and timing of wastewater discharges, potentially overwhelming treatment systems during storm events or causing water scarcity during dry periods (Liu et al. 2013; Escobar-Camacho et al. 2021; Dueñas et al. 2021). In addition, rising sea levels associated with climate change and extreme weather events can increase the risk of coastal flooding, contamination of water bodies, sewage overflows, and saltwater intrusion into wastewater infrastructure (Gupta et al. 2021; Shaw and Dorea 2021).

Improving wastewater treatment is essential for the long-term environmental sustainability of the Galápagos Islands. Examples of alternative successful approaches include advanced technologies such as membrane bioreactors used in Singapore, and decentralized systems like constructed wetlands employed in the

Maldives. Hawaii utilizes UV disinfection, while Barbados invests in treatment plants based on biological processes. Palau enforces strict regulations and promotes eco-friendly technologies, including composting toilets. In Haiti, container-based sanitation has been implemented as a cost-effective solution (Tilmans et al. 2015; Russel et al. 2015; Xiang et al. 2020; Sewak et al. 2021; VanRiper et al. 2022; Yeneceri and Tortum Yeneceri 2023; Khalid Mohamed et al. 2023).

These approaches by small island developing states (SIDS), which, like other SIDS, including the Galápagos Islands, face unique challenges such as limited land area, dense populations, and reliance on tourism. It requires that the Galápagos Island's fragile ecosystems, which are critical for biodiversity and global conservation, demand tailored solutions to protect marine ecosystems and public health while promoting sustainable development. The Galápagos Islands can mitigate pollution, conserve water resources, and preserve its fragile coastal environments by implementing these strategies. These successful examples demonstrate the importance of proactive measures in wastewater management, ensuring the long-term well-being of the island's human communities and its world-renowned ecosystems.

Governance, Policy, Regulations, and Stakeholders' Engagement

Safe water can be provided through the feasible management of water supply and sanitation systems, as stated in the United Nations Sustainable Development Goal 6 (SDG 6). To achieve this, a safe water source must be accessible, available, and free from contamination (Grube et al. 2020). Access to water is a constitutional right in Ecuador and represents a fundamental pillar of societal equality. However, despite this legal framework, geographic and economic discrepancies have deepened disparities among Ecuador's populace. This divide is particularly pronounced in the Galápagos Islands, where unique challenges underscore the struggle to enforce regulations about water management (Mateus et al. 2020; Wingfield et al. 2021; Mateus and Quiroga 2022; Vásquez et al. 2022).

At the heart of these challenges lie the following main issues: water scarcity, water quality, inadequate wastewater treatment, and the absence of adequate water reuse mechanisms. The fragile ecosystem of the Galápagos amplifies these obstacles, making sustainable water management an urgent priority. Addressing these challenges demands innovative solutions and a concerted effort to bridge the gaps in access and infrastructure across the islands. Only through collaborative action can the promise of equal access to water be achieved for all inhabitants of the Galápagos and beyond (Mateus et al. 2020).

Although a substantial investment of resources has been directed toward water management initiatives in the Galápagos Islands, the anticipated improvements have yet to materialize. This shortfall highlights critical issues such as a lack of transparency and accountability in project implementation (Mateus et al. 2020;

Vásquez 2022). To address these challenges, establishing mechanisms that ensure tangible and measurable outcomes is essential. These mechanisms could include stakeholder collaboration, rigorous monitoring and evaluation systems, and innovative funding structures to drive sustainable water management solutions in the Islands.

Against this backdrop of complex challenges in water management, it becomes increasingly evident that community engagement and stakeholder involvement are essential components of any effective solution. Without the active participation of local communities and key stakeholders, efforts to address water scarcity and quality issues are likely to fall short (Mateus et al. 2020).

One approach to foster community engagement is the implementation of user fees (Wingfield et al. 2021; Re et al. 2023). By introducing a system where users pay for the water they consume, individuals become more conscious of their usage and are incentivized to adopt water-saving practices. This helps conserve water resources and generates revenue that can be reinvested into improving infrastructure and implementing sustainable water management practices (Grube et al. 2020; Re et al. 2023). For instance, households in Santa Cruz Island are willing to pay approximately $36/month, about 2.1% of average monthly household income, for concurrent system reliability and water quality improvements. This willingness to pay contrasts with a flat rate structure not contingent on water consumption (Vásquez et al. 2021).

Additionally, awareness campaigns play a crucial role in educating the public about the importance of water conservation and the impact of their actions on the local water ecosystem. Through targeted messaging and educational programs, communities can be empowered to make informed decisions about their water usage, leading to more responsible behavior and a greater sense of ownership over water resources (Mateus et al. 2020).

Fostering active participation involves creating platforms for dialogue and collaboration between various stakeholders, including government agencies, non-governmental organizations (NGOs), businesses, and local communities. By involving stakeholders in decision-making processes and encouraging collaboration, effective solutions and inclusive of diverse perspectives and needs can be developed.

Implementing regulations on the Galápagos Islands encounters logistical difficulties due to their remote location and limited resources, leading to challenges in monitoring and enforcement (Jones 2013; Burbano et al. 2020; Alava et al. 2023). Balancing tourism development with conservation goals is complex, and better governance is needed for effective coordination (Burbano and Meredith 2021; Vásquez 2022). Such circumstances make community involvement and stakeholder engagement essential by avoiding the exclusionary decision-making processes (Hoyman and McCall 2013; Burbano et al. 2020). Local communities contribute valuable knowledge and expertise, fostering a sense of ownership over natural resources, empowering them with alternative livelihoods promotes sustainable practices and social cohesion (Hoyman and McCall 2013). Collaborative partnerships facilitate collective action, while involving locals in monitoring efforts enhances data collection and supports conservation strategies (Hoyman and McCall 2013; Jones 2013). User payment schemes also incentivize responsible resource use and support conservation initiatives financially (Tilmans et al. 2014).

The successful management of water resources requires a holistic and collaborative approach that addresses the ecological, social, and economic dimensions of water sustainability (Mateus et al. 2020). Central to this effort is the principle of environmental justice, which ensures that all communities, particularly marginalized and vulnerable populations, have equitable access to clean water and an active role in resource management. Given that freshwater is both scarce and finite in the Galápagos Archipelago, it is imperative to conserve natural watersheds, promote sustainable water use practices, adopt integrated water resources management approaches, install wastewater treatment systems, build climate resilience, and empower local communities. By doing som island ecosystems can thrive in the face of water scarcity, wastewater management, and climate change all while ensuring a just and sustainable future for both people and nature (Vásquez et al. 2021, 2022; Mateus and Quiroga 2022).

Current Efforts

Given the importance of managing freshwater sources, water quality, public health, and wastewater treatment in the Galápagos Islands, various initiatives currently focus on conservation and innovative technologies to address freshwater needs and wastewater management while minimizing environmental impact.

Freshwater Management

In recent years, efforts have been made to address water source management challenges in the Galápagos Islands. These efforts encompass a range of initiatives, including conservation measures, technological innovations, community engagement, and policy interventions.

(a) *Rainwater/fog harvesting and storage* systems implemented in rural areas of the Galápagos Islands can supplement freshwater supplies during the dry season. This is feasible due to fog prevalence on the islands for approximately six months each year (Schmitt et al. 2018). For instance, two standard fog collectors have been installed at an elevation of 600 m on San Cristóbal Island. These collectors gather water, which is stored in cisterns or tanks for household use. This strategy significantly reduced reliance on groundwater and surface water sources. Results indicated that fog collectors, with an area of 0.15 m^2 each, were able to supply up to 25% of the water deficit during a dry year (Echeverría et al. 2020).

In addition, Galápagos Resiliente (https://Galápagosresiliente.weebly.com/) reported that 6.9% and 4.6% of water needs are supplied from rainwater collection in Santa Cruz and Isabela, respectively.

(b) *Desalination plants* have been employed to produce potable water from seawater or brackish groundwater in the Galápagos Islands, both in urban centers (Santa Cruz and Isabela) and in tourist facilities and ships. The desalination plant in Santa Cruz is expected to be administered by the municipality, and it should meet the entire demand of the public network by the end of this year. In Isabela, a desalination plant has been operational since 2014. Additionally, Floreana also has small desalination plants to preserve surface reservoirs. However, these technologies present several logistical and power requirements limitations because the energy provided for brackish water pumping and desalination units is generated from fossil fuels. In addition, some issues from the dumping of brine have been identified as potential environmental impacts of this technology (Vásquez et al. 2022). Currently, new sustainable energy sources based on photovoltaic generation are being evaluated for application in desalination systems on Santa Cruz and Floreana islands (Amoroso et al. 2021; Arévalo et al. 2022).

(c) *Groundwater Management* and monitoring in the Galápagos islands is crucial to prevent over-extraction and seawater intrusion (Riascos-Flores et al. 2021). For instance, the municipality of Santa Cruz has investigated the "El Gallito" a freshwater source as an alternative water supply that allows water systems to reduce their dependence on aquifers by 10%, and in the event of an emergency, they can provide a minimum amount of vital water for the population (https://www.gwp.org/es/agentes-de-cambio-por-el-agua/change-stories/559192/). Moreover, groundwater contamination resulting from anthropogenic activities, such as agriculture and pollution, poses significant risks to water quality and public health (Mateus et al. 2019; Fernandes Reyes 2022).

(d) *Water Conservation and Efficiency* measures have been promoted in the Galápagos Islands to reduce water demand and minimize wastage. Public awareness campaigns, educational programs, and water-saving technologies could encourage residents and businesses to adopt practices that conserve water, such as fixing leaks, using water-efficient appliances, and practicing water-wise landscaping (Re et al. 2023).

(e) *Ecosystem Restoration and Conservation* efforts are underway to protect the Galápagos Islands' natural watersheds, wetlands, and riparian zones. These efforts are being conducted in Floreana and 16 other sites around the Galápagos Islands, aimed at preserving critical habitat for native species, maintaining water quality, and regulating water flow, contributing to the overall health and resilience of freshwater ecosystems (Negoita et al. 2022; Rhea et al. 2023).

(f) *Community Engagement and Empowerment.* Participatory approaches, community-based monitoring, and the integration of traditional knowledge enhance social cohesion, resilience, and equitable access to water resources in island ecosystems (Re et al. 2023). Academic institutions, the Galápagos National Park, and the local government have promoted community engagement by conducting seminars and communication campaigns to educate and raise awareness about water conservation and quality.

(g) *Policy and governance are key to the Galápagos Islands.* Government agencies, NGOs, and community groups have collaborated to develop and implement policies and regulations that promote sustainable water management in the Galápagos Islands. These policies address issues such as water allocation, pollution control, land use planning, and climate change adaptation, providing a framework for integrated water resources management (Wingfield et al. 2021; Alava et al. 2023). Studies have been performed to assess commitment in the decision-making of the community in the Galápagos (Houck et al. 2020; Mateus et al. 2020).

(h) *Research and Monitoring* activities have been conducted to assess the status of freshwater resources in the Galápagos Islands and identify emerging challenges and opportunities. Scientific studies, data collection efforts, and environmental monitoring programs provide valuable insights into water availability, quality, and ecosystem health, informing evidence-based decision-making and adaptive management strategies (Re et al. 2023). Collaborative research efforts have brought together experts from diverse disciplines, including hydrology, ecology, geology, climatology, social sciences, and engineering, to study various aspects of water management in the Galápagos Islands. By integrating multiple perspectives and methodologies, interdisciplinary teams can provide comprehensive insights into the complex interactions between natural and human systems. In addition, collaborations with international research institutions, universities, and funding agencies offer opportunities for knowledge exchange, capacity-building, and access to expertise and resources from around the world. For instance, the long-term water quality monitoring program developed by the Galápagos Science Center with the participation of the Galápagos National Park, the Municipality of San Cristóbal, and the local community represents the most extensive longitudinal study of water quality across natural and engineered systems in the Galápagos Islands (Grube et al. 2020).

Wastewater Management

Wastewater management across the Galápagos Archipelago varies greatly, with most wastewater remaining untreated and only minimal efforts undertaken to address the issue effectively:

(a) *Wastewater treatment plants.* The wastewater treatment plant in San Cristóbal, the biggest in Galápagos, is based on conventional technologies. Nevertheless, the operation of this system is constrained by a range of factors, including significant energy inputs, exceeding treatment capacities, reliance on infrastructure such as settling tanks, aeration basins, and clarifiers, substantial capital and operational costs, the need for skilled labor, and logistical challenges. Hence, implementing sustainable wastewater management practices, enhancing sanitation infrastructure, and fostering international collaboration are not just strategic choices but imperatives in safeguarding health at the Galápagos Islands' local, global, and ecological levels (Nicholas et al. 2020). The implications of wastewater on Public Health, Global Health, and One Health demand integrated and collaborative efforts (Nicholas et al. 2020; Grube et al. 2020; Thompson et al. 2023).

(b) *Nature-based solutions* such as constructed wetlands are engineered systems designed to replicate the functions of natural wetlands for wastewater treatment using native aquatic plants. The treated effluent can then be discharged to surface water bodies, used for irrigation, or further treated for reuse purposes (Vymazal et al. 2021). Since 2016, a treatment system on Isabela has been operated to treat wastewater generated by its inhabitants. This constructed wetland occupies an area of 6600 m^2 (https://www.wwf.org.ec/?264791/El-tratamiento-biolgico-de-aguas-residuales-fundamento-de-una-comunidad-saludable).

(c) *Septic tanks* are widely used in Santa Cruz and some rural areas, where connection to a centralized sewerage system is unavailable. Improperly maintained septic tanks can pose environmental and public health risks, as is evidenced in some areas in Santa Cruz (Mateus et al. 2019).

(d) *Research* studies on wastewater management in the Galápagos Islands are scarce, with existing studies primarily focusing on the management of urban wastewater in Santa Cruz Island and the impacts of anthropogenic pollution from tourism, agriculture and urbanization on water quality in Santa Cruz, Floreana, and Isabela (Ragazzi et al. 2016; Riascos-Flores et al. 2025). As an alternative to ineffective septic tanks, a modular, easily replicable and low-maintenance wastewater treatment system has been proposed for Santa Cruz Island, offering an innovative solution. Sewage has been identified as a significant contributor to nitrate contamination across Santa Cruz, Isabela, and Floreana. Fecal coliforms have been detected in coastal urban waters and water sources used for treatment, highlighting a persistent contamination cycle. These studies emphasize the critical need for improved sewage and agricultural management in the Galápagos Islands to enhance and protect the Islands' highly vulnerable ecosystems and the vital water provisioning services they support.

Lessons Learned and Recommendations

Advanced Research and Ecosystem Understanding: Fresh- and brackish-water ecosystems in the Galápagos Islands remain understudied, with limited knowledge of their functioning, biodiversity, and ecosystem services. Future research should prioritize exploring their diversity and endemism as indicators of ecosystem health, along with developing long-term biological monitoring programs to detect and address negative impacts. Additionally, research must focus on climate change adaptation and mitigation strategies, acknowledging these ecosystems' critical role in supporting endemic biodiversity, enhancing island resilience, and contributing to conservation.

Integrated Water Management Approaches: Lessons from existing water management efforts in the Galápagos highlight the importance of a holistic approach that recognizes freshwater as a scarce and finite resource. This approach should consider the entire water cycle—including quality assurance, wastewater treatment, reuse, and ecosystem protection—to optimize resource use, minimize environmental impacts, and enhance climate resilience.

Decentralized and Nature-Based Solutions: The majority of wastewater in the Galápagos is discharged untreated into the marine environment. Given the unique population distribution and topography of the Archipelago, decentralized wastewater treatment systems are promising. These systems offer scalability, cost-effectiveness, and flexibility, particularly in remote areas, and benefit from advances in sensor technology, automation, and data analytics for real-time monitoring and predictive maintenance. Nature-based solutions, such as constructed wetlands and green infrastructure, can further address wastewater management challenges by leveraging natural processes to treat water, enhance resilience, and mitigate climate impacts.

Treated Wastewater Reuse and Resource Recovery: No treated wastewater reuse or resource recovery currently takes place in the Galápagos. Implementing reuse of treated wastewater for irrigation, groundwater recharge, and non-potable purposes could significantly reduce pressure on freshwater supplies and promote sustainable resource conservation. Additionally, recovering nutrients for agriculture and energy from biogas will advance sustainable development and foster a circular economy, while reducing dependency on external inputs. Furthermore, monitoring emerging contaminants is critical to protect the fragile ecosystems of the Galápagos.

Community Engagement and Governance: Community members and local stakeholders are frequently excluded from decision-making and enforcement processes related to water management in the Galápagos. Actively engaging these communities fosters ownership, builds trust, and improves the social acceptability of projects across the Archipelago, while also promoting environmental justice by ensuring that the rights and needs of all stakeholders are respected. It is therefore recommended to implement participatory approaches that reflect local values and perspectives, supported by strong accountability mechanisms and transparent policies, to ensure the equitable and successful adoption of sustainable solutions.

Global Implications and Transferable Practices: The Galápagos Islands' experience provides valuable insights for other islands and regions facing similar challenges. Collaboration among stakeholders—government, industry, academia, and communities—is essential to developing adaptive, integrated strategies. Incorporating climate change scenarios, assessing vulnerability, and designing flexible infrastructure are critical to enhancing resilience.

Addressing Challenges and Ensuring Outcomes: Despite significant investments in water management, progress in the Galápagos has been limited due to issues such as transparency and accountability in project implementation. Establishing mechanisms for measurable outcomes, rigorous monitoring, and innovative funding structures is critical to achieving sustainable water management in the Archipelago.

Conclusions

Challenges of Limited Freshwater Resources

The Galápagos Islands face significant challenges due to limited freshwater resources. Water availability and treatment processes are severely constrained, optimizing water use, exploring sustainable alternatives, and minimizing losses throughout treatment and distribution is crucial. These challenges are further exacerbated by population growth and the pressures of tourism, which significantly increase demand and strain existing systems. Addressing these issues requires not only technical solutions but also a bioethical framework that considers the equitable distribution of resources and the protection of the islands' unique ecosystems. Conservation measures such as promoting efficient water usage among residents and tourists, adopting advanced treatment technologies like desalination powered by renewable energy, and utilizing rainwater harvesting can provide sustainable solutions. Additionally, addressing inefficiencies in water systems by modernizing infrastructure and reducing leaks is essential to ensure that the limited water resources are used effectively.

Pressures on Water Infrastructure and Resources

Rapid population growth and a surge in tourism have placed immense pressure on the Galápagos Islands' water infrastructure and resources. The increasing demand for water and sanitation services necessitates careful planning and investment. Integrated water management is essential to balance the needs of residents, ecosystems, and tourists. Limiting tourist numbers, enforcing eco-tourism practices, and ensuring that tourism infrastructure adheres to environmental standards can alleviate the resource strain. Expanding and upgrading water and wastewater systems is critical to accommodate the growing population while safeguarding environmental and public health.

Environmental Impacts and Ecosystem Sensitivity

The islands' fragile ecosystems are susceptible to water usage, pollution, and environmental degradation which threaten their resilience and risk pushing them toward critical tipping points. Excessive pumping of aquifers and the discharge of untreated wastewater profoundly impact marine and terrestrial ecosystems, including coral reefs, coastal habitats, and freshwater sources. Protecting these ecosystems requires strict monitoring and enforcement of environmental standards, alongside investments in ecosystem restoration projects, such as mangrove rehabilitation. Engaging local communities in conservation efforts and promoting sustainable practices can further protect the delicate balance of these environments while fostering local stewardship and just resource management.

Constraints on Wastewater Infrastructure Expansion

Land availability on the islands is limited, creating challenges for the siting and expansion of wastewater treatment infrastructure. To address this constraint, innovative solutions—including nature-based approaches and green technologies—play a vital role. Compact and space-efficient treatment systems such as membrane bioreactors or vertical treatment setups minimize physical footprints, while decentralized wastewater management allows for localized, flexible treatment options that reduce reliance on large-scale infrastructure. Integrating nature-based solutions like constructed wetlands or biofiltration not only enhances treatment efficiency but also supports ecosystem health. For sustainable growth, future development plans must prioritize land allocation for critical infrastructure while emphasizing these environmentally friendly and space-saving technologies.

Energy and Sustainable Challenges

Energy demands for tap water and wastewater treatment further complicate the situation, as the islands rely heavily on imported fossil fuels. This dependence makes the islands vulnerable to energy price fluctuations and supply disruptions. To enhance sustainability, adopting energy-efficient treatment technologies, such as anaerobic digestion or constructed wetlands, and integrating renewable energy sources like solar, wind, or geothermal power into water management systems is vital. Additionally, recovering energy from treatment processes, such as generating biogas through anaerobic digestion of wastewater, can offset energy demands and reduce reliance on external resources. Furthermore, addressing the complex interconnections between biodiversity, water, energy, and food—the B-WEF nexus—should be prioritized to ensure holistic and sustainable management within this unique ecosystem.

Impacts of Climate Change

Climate change compounds these challenges, with rising sea levels, changes in precipitation patterns, and more frequent extreme weather events posing significant risks to water management systems. Saltwater intrusion, flooding, and infrastructure damage are likely to compromise the effectiveness and resilience of these systems. Adaptation measures, such as constructing flood-resistant infrastructure and expanding water storage capacities, are crucial for mitigating these risks. Additionally, restoring wetlands in the Islands offers a multifaceted solution. Wetlands act as natural buffers against flooding, improve water quality, and sequester carbon dioxide, thereby reducing greenhouse gas emissions. These restoration efforts also bolster ecosystem health, enhance biodiversity, and contribute to climate change mitigation. Complementary strategies, such as establishing climate monitoring and early warning systems, further strengthen preparedness and resilience against climate-related disruptions.

The Need for a Multidisciplinary Approach

Addressing the water-related challenges in the Galápagos Islands—from water scarcity to wastewater management—demands a multifaceted and multidisciplinary approach. This approach must integrate environmental science, biology, hydrology, engineering, and public health. Special emphasis must be placed on recognizing the limits of freshwater availability in the Galápagos, acknowledging its scarcity and finite nature, and ensuring water management operates within the boundaries of this highly sensitive and ecologically unique island system. This framework underscores the critical role of freshwater ecosystems in supporting biodiversity, enhancing island resilience, and promoting conservation. By adopting a holistic perspective, their value is seen as extending beyond meeting human needs, emphasizing the importance of balanced and sustainable approaches to their protection and stewardship. Furthermore, active stakeholder engagement and the implementation of robust policies and regulations are essential to ensure that proposed solutions are not only scientifically grounded but also practical and contextually sustainable for the Galápagos. These efforts must prioritize environmental justice, safeguarding sustainable water governance that upholds the rights and well-being of current and future generations while fostering long-term resilience and sustainability within this fragile island ecosystem.

References

Alava JJ, McMullen K, Jones J et al (2023) Multiple anthropogenic stressors in the Galápagos Islands' complex social–ecological system: interactions of marine pollution, fishing pressure, and climate change with management recommendations. Integr Environ Assess Manag 19:870–895. https://doi.org/10.1002/ieam.4661

Alomía Herrera I, Paque R, Maertens M, Vanacker V (2022) History of land cover change on Santa Cruz Island, Galápagos. Land (Basel) 11:1017. https://doi.org/10.3390/land11071017

Amoroso F, Hidalgo-León R, Litardo J et al (2021) Simulations of solar power systems to provide electricity to a model water desalination Plant in Floreana Island, Ecuador. In: ASME 2021 15th international conference on energy sustainability. American Society of Mechanical Engineers

Arévalo P, Eras-Almeida AA, Cano A et al (2022) Planning of electrical energy for the Galápagos Islands using different renewable energy technologies. Electr Power Syst Res 203:107660. https://doi.org/10.1016/j.epsr.2021.107660

Auken E, Violette S, d'Ozouville N et al (2009) An integrated study of the hydrogeology of volcanic islands using helicopter borne transient electromagnetic: application in the Galápagos archipelago. Comptes Rendus Géoscience 341:899–907. https://doi.org/10.1016/j.crte.2009.07.006

Ayala-Pico J, Arcos-Aviles D, Ibarra A et al (2023) Current development of electricity generation systems in the Galápagos Islands – Ecuador. Renewable Energy Focus 46:88–102. https://doi.org/10.1016/j.ref.2023.06.003

Badhwa N, Fejfar D, Pozo R, Nicholas K, Grube A, Stewart J, Thompson A, Ochoa-Herrera V (2022) Water quality and access in Isabela: results from a household water survey. In: Water, food and human health in the Galapagos, Ecuador: "A little world within itself". Springer International Publishing, Cham, pp 57–74

Bak M, Kociolek JP, Lange-Bertalot H, Lopato D, Witkowski A (2017) Novel diatom species (Bacillariophyta) from the freshwater discharge site of Laguna Diablas (Island Isabela= Albemarle) from the Galapagos. Phytotaxa 311(3):201–224

Barragán KS, Rosero-López D, Cordero-Rivera A, Escobar-Camacho D, Andrade S, Moreno P, Daza J, Encalada C, A. (2025) Galápagos invertebrados acuáticos, aquatic invertebrates. USFQ PRESS

Burbano DV, Meredith TC (2021) Effects of tourism growth in a UNESCO world heritage site: resource-based livelihood diversification in the Galápagos Islands, Ecuador. J Sustain Tour 29:1270–1289. https://doi.org/10.1080/09669582.2020.1832101

Burbano DV, Meredith TC, Mulrennan ME (2020) Exclusionary decision-making processes in marine governance: the rezoning plan for the protected areas of the 'iconic' Galápagos Islands, Ecuador. Ocean Coast Manag 185:105066. https://doi.org/10.1016/j.ocecoaman.2019.105066

Bush MB, Steinitz-Kannan M, Sachs JP, Cole JE, Collins AF, Conroy JL, Restrepo A, Zhang Z (2010) Forty years of paleoecology in the galápagos

Cárdenas López P (2019) Composición del Archipiélago de Galápagos. Invertebrados acuáticos. Tesis (Bióloga), Universidad San Francisco de Quito, Colegio de Ciencias Biológicas y Ambientales; Quito, Ecuador. https://repositorio.usfq.edu.ec/handle/23000/10087

Causton CE, Peck SB, Sinclair BJ, Roque-Albelo L, Hodgson CJ, Landry B (2006) Alien insects: threats and implications for conservation of Galápagos Islands. Ann Entomol Soc Am 99(1):121–143

Cordovez V, Carrion V, Torres G, Ortiz D, Cabrera W, Balian H, Vivanco A, Pérez Jaramillo J, Chaves J, Pazmino D, Hof P, Raaijmakers J (2024) Darwin's expedition revisited to reveal the evolution of plant–microbe interactions on Galápagos. Nat Microbiol 9. https://doi.org/10.1038/s41564-024-01749-z

Dueñas A, Jiménez-Uzcátegui G, Bosker T (2021) The effects of climate change on wildlife biodiversity of the Galápagos islands. Climate Change Ecology 2:100026. https://doi.org/10.1016/j.ecochg.2021.100026

Echelpoel F et al (2019) Spatial characteristics and temporal evolution of chemical and biological freshwater status as baseline assessment on the Tropical Island San Cristóbal (Galápagos, Ecuador). Water (Basel) 11:880. https://doi.org/10.3390/w11050880
Echeverría P, Domínguez C, Villacís M, Violette S (2020) Fog harvesting potential for domestic rural use and irrigation in San Cristóbal Island, Galápagos, Ecuador. Cuadernos de Investigación Geográfica 46:563–580. https://doi.org/10.18172/cig.4382
Escobar-Camacho D, Rosero P, Castrejón M et al (2021) Oceanic islands and climate: using a multi-criteria model of drivers of change to select key conservation areas in Galápagos. Reg Environ Chang 21:47. https://doi.org/10.1007/s10113-021-01768-0
Fernandes Reyes M (2022) Analysis of water supply and demand management strategies in overcoming water scarcity in Santa Cruz. In: Water, food and human health in the Galapagos, Ecuador: "A little world within itself". Springer International Publishing, Cham, pp 9–27
Fernandes Reyes MFR, Trifunović N, Sharma S et al (2017) Quantification of urban water demand in the Island of Santa Cruz (Galápagos archipelago). Desalination Water Treat 64:1–11. https://doi.org/10.5004/dwt.2017.20284
Gerecke R, Peck SB, Pehofer HE (1995) The invertebrate fauna of the inland waters of the Galápagos archipelago (Ecuador): a limnological and zoogeographical summary. Archiv für Hydrobiologie Supplementband, Algological studies 107(2):113–147
Gough D, Oliver S, Thomas J (2017) An introduction to systematic reviews, 2nd edn
Grube AM, Stewart JR, Ochoa-Herrera V (2020) The challenge of achieving safely managed drinking water supply on San Cristóbal Island, Galápagos. Int J Hyg Environ Health 228:113547. https://doi.org/10.1016/j.ijheh.2020.113547
Gupta S, Mittal Y, Panja R et al (2021) Conventional wastewater treatment technologies. In: Current developments in biotechnology and bioengineering. Elsevier, pp 47–75
He C, Liu Z, Wu J et al (2021) Future global urban water scarcity and potential solutions. Nat Commun 12:4667. https://doi.org/10.1038/s41467-021-25026-3
Heads M, Grehan JR (2021) The Galápagos islands: biogeographic patterns and geology. Biol Rev 96:1160–1185. https://doi.org/10.1111/brv.12696
Houck KM, Terán E, Ochoa J et al (2020) Drinking water improvements and rates of urinary and gastrointestinal infections in Galápagos, Ecuador: assessing household and community factors. Am J Hum Biol 32. https://doi.org/10.1002/ajhb.23358
Hoyman MM, McCall JR (2013) Is there trouble in paradise? The perspectives of Galápagos community leaders on managing economic development and environmental conservation through ecotourism policies and the special law of 1998. J Ecotour 12:33–48. https://doi.org/10.1080/14724049.2012.749882
INEC (2022) Ecuador en Cifras. https://www.ecuadorencifras.gob.ec/censo-de-poblacion-y-vivienda-Galápagos/
Jones PJS (2013) A governance analysis of the Galápagos marine reserve. Mar Policy 41:65–71. https://doi.org/10.1016/j.marpol.2012.12.019
Karnauskas KB, Murtugudde R, Owens WB (2014) Climate and the global reach of the Galápagos archipelago: state of the knowledge. In: The Galápagos: A natural laboratory for the earth sciences, pp 215–231
Khalid Mohamed S, Hernandez HG, Hooijmans CM (2023) Technology selection for wastewater treatment in the Maldives. Maldives Natl J Res 11:56–76. https://doi.org/10.62338/4bbe9878
Liu Y, Xie L, Morrison JM, Kamykowski D (2013) Dynamic downscaling of the impact of climate change on the ocean circulation in the Galápagos archipelago. Adv Meteorol 2013:1–18. https://doi.org/10.1155/2013/837432
Mateus C, Quiroga D (2022) Galapagos' water management evaluation under a changing climate and the current COVID-19 pandemic. In: Water, food and human health in the Galapagos, Ecuador: "A little world within itself". Springer International Publishing, Cham, pp 29–55
Mateus C, Guerrero CA, Quezada G et al (2019) An integrated approach for evaluating water quality between 2007–2015 in Santa Cruz Island in the Galápagos archipelago. Water (Basel) 11:937. https://doi.org/10.3390/w11050937

Mateus C, Valencia M, DiFrancesco K et al (2020) Governance mechanisms and barriers for achieving water quality improvements in Galápagos. Sustainability 12:8851. https://doi.org/10.3390/su12218851

Mena CF, Quiroga D, Walsh SJ (2020) Threats to sustainability in the Galápagos Islands: a social–ecological perspective. In: The Elgar companion to geography, Transdisciplinarity and sustainability. Edward Elgar Publishing

Miller J, Adair L, Stewart J et al (2022) Water insecurity is differentially associated with food insecurity across seasons: evidence from the Galápagos. Curr Dev Nutr 6:591. https://doi.org/10.1093/cdn/nzac060.049

Mosquera JD, Valle CA, Nieto-Claudin A et al (2023) Prevalence of toxoplasma gondii in Galápagos birds: inference of risk factors associated with diet. PLoS One 18:e0287403. https://doi.org/10.1371/journal.pone.0287403

Negoita L, Gibbs JP, Jaramillo Díaz P (2022) Cost-effectiveness of water-saving technologies for restoration of tropical dry forest: a case study from the Galápagos Islands, Ecuador. Restor Ecol 30. https://doi.org/10.1111/rec.13576

Nicholas K, Bentley M, Terán E, Thompson A (2020) Water security in the Galápagos: socioecological determinants and health implications. EcoHealth 17:111–124. https://doi.org/10.1007/s10393-019-01456-w

Paltán HA, Benitez FL, Rosero P et al (2021) Climate and sea surface trends in the Galápagos Islands. Sci Rep 11:14465. https://doi.org/10.1038/s41598-021-93870-w

Paltán HA, Benítez FL, Narvaez M et al (2023) Water security and agricultural systems in the Galápagos Islands: vulnerabilities under uncertain future climate and land use pathways. Front Water 5. https://doi.org/10.3389/frwa.2023.1245207

Percy MS, Schmitt SR, Riveros-Iregui DA, Mirus BB (2016) The Galápagos archipelago: a natural laboratory to examine sharp hydroclimatic, geologic and anthropogenic gradients. WIREs Water 3:587–600. https://doi.org/10.1002/wat2.1145

Pryet A, Domínguez C, Tomai PF et al (2012) Quantification of cloud water interception along the windward slope of Santa Cruz Island, Galápagos (Ecuador). Agric Meteorol 161:94–106. https://doi.org/10.1016/j.agrformet.2012.03.018

Ragazzi M, Catellani R, Rada E et al (2016) Management of urban wastewater on one of the Galápagos Islands. Sustainability 8:208. https://doi.org/10.3390/su8030208

Re V, Rizzi J, Tuci C et al (2023) Challenges and opportunities of water quality monitoring and multi-stakeholder management in small islands: the case of Santa Cruz, Galápagos (Ecuador). Environ Dev Sustain 25:3867–3891. https://doi.org/10.1007/s10668-022-02219-4

Reyes M, Trifunović N, Sharma S et al (2017) Mitigation options for future water scarcity: a case study in Santa Cruz Island (Galápagos archipelago). Water (Basel) 9:597. https://doi.org/10.3390/w9080597

Rhea S, Camacho BE, Amoriello CW et al (2023) Assessing livestock production practices on small-scale multi-species farms located on Floreana Island, Galápagos Islands. Animals 13:686. https://doi.org/10.3390/ani13040686

Riascos-Flores L, Bruneel S, Van der Heyden C et al (2021) Polluted paradise: occurrence of pesticide residues within the urban coastal zones of Santa Cruz and Isabela (Galápagos, Ecuador). Sci Total Environ 763:142956. https://doi.org/10.1016/j.scitotenv.2020.142956

Riascos-Flores L, Ho L, Van Echelpoel W et al (2025) Chemical and microbiological analysis of urban and associated natural water systems of inhabited volcanic islands of the Galápagos (Ecuador). Water Res 281:123516. https://doi.org/10.1016/j.watres.2025.123516

Russel K, Tilmans S, Kramer S et al (2015) User perceptions of and willingness to pay for household container-based sanitation services: experience from cap Haitien, Haiti. Environ Urban 27:525–540. https://doi.org/10.1177/0956247815596522

Schmitt SR, Riveros-Iregui DA, Hu J (2018) The role of fog, orography, and seasonality on precipitation in a semiarid, tropical island. Hydrol Process 32:2792–2805. https://doi.org/10.1002/hyp.13228

Sewak A, Deshpande S, Rundle-Thiele S et al (2021) Community perspectives and engagement in sustainable solid waste management (SWM) in Fiji: a socioecological thematic analysis. J Environ Manag 298:113455. https://doi.org/10.1016/j.jenvman.2021.113455

Shaw K, Dorea CC (2021) Biodegradation mechanisms and functional microbiology in conventional septic tanks: a systematic review and meta-analysis. Environ Sci (Camb) 7:144–155. https://doi.org/10.1039/D0EW00795A

Thompson AL, Nicholas KM, Watson E et al (2020) Water, food, and the dual burden of disease in Galápagos, Ecuador. Am J Hum Biol 32. https://doi.org/10.1002/ajhb.23344

Thompson AL, Stewart JR, Bentley M, Ocampo Trujillo JE, Teran E, Ochoa V (2023) Climate and health challenges in small Island states: identifying vulnerability in water and food resources in the Galapagos Islands, Ecuador. In: Island ecosystems: challenges to sustainability. Springer International Publishing, Cham, pp 91–105

Tilmans S, Diaz-Hernandez A, Nyman E, Davis J (2014) The potential for financing small-scale wastewater treatment through resource recovery: experience from Bocas del Toro, Panama. J Water Sanit Hyg Dev 4:449–459. https://doi.org/10.2166/washdev.2014.138

Tilmans S, Russel K, Sklar R et al (2015) Container-based sanitation: assessing costs and effectiveness of excreta management in Cap Haitien, Haiti. Environ Urban 27:89–104. https://doi.org/10.1177/0956247815572746

VanRiper F, Russel KC, Cramer LA et al (2022) Container-based sanitation services and attrition: an examination of drivers and implications. Front Environ Sci 9. https://doi.org/10.3389/fenvs.2021.817142

Vásquez WF (2022) Tourists' willingness to fund improvements of local water services: a contingent valuation study from the Galapagos. In: Water, food and human health in the Galapagos, Ecuador: "A little world within itself". Springer International Publishing, Cham, pp 75–87

Vásquez WF, Raheem N, Quiroga D, Ochoa-Herrera V (2021) Household preferences for improved water services in the Galápagos Islands. Water Resour Econ 34:100180. https://doi.org/10.1016/j.wre.2021.100180

Vásquez WF, Raheem N, Quiroga D, Ochoa-Herrera V (2022) Valuing improved water services and negative environmental externalities from seawater desalination technology: a choice experiment from the Galápagos. J Environ Manag 304:114204. https://doi.org/10.1016/j.jenvman.2021.114204

Violette S, d'Ozouville N, Pryet A, Deffontaines B, Fortin J, Adelinet M (2014) Hydrogeology of the Galapagos Archipelago: an integrated and comparative approach between islands. In: The Galapagos: a natural laboratory for the Earth sciences, pp 167–183

Vymazal J, Zhao Y, Mander Ü (2021) Recent research challenges in constructed wetlands for wastewater treatment: a review. Ecol Eng 169:106318. https://doi.org/10.1016/j.ecoleng.2021.106318

Walsh SJ, McCleary AL, Heumann BW et al (2010) Community expansion and infrastructure development: implications for human health and environmental quality in the Galápagos Islands of Ecuador. J Lat Am Geogr 9:137–159. https://doi.org/10.1353/lag.2010.0024

Wingfield S, Martínez-Moscoso A, Quiroga D, Ochoa-Herrera V (2021) Challenges to water management in Ecuador: legal authorization, quality parameters, and socio-political responses. Water (Basel) 13:1017. https://doi.org/10.3390/w13081017

Xiang Y, Gonsior M, Schmitt-Kopplin P, Shang C (2020) Influence of the UV/H_2O_2 advanced oxidation process on dissolved organic matter and the connection between elemental composition and disinfection Byproduct formation. Environ Sci Technol 54:14964–14973. https://doi.org/10.1021/acs.est.0c03220

Yeneceri M, Tortum Yeneceri K (2023) Recycling wastewater with membrane technology and the case of Singapore. Bitlis Eren Üniversitesi Fen Bilimleri Dergisi 12:478–484. https://doi.org/10.17798/bitlisfen.1242232

Chapter 6
Energy Transition and Climate Action in Galapagos: Policy Evaluation and Guiding Principles

Andreanne Doyon, María Isabel Loyola Plúa, Chingwen Cheng, and Yekang Ko

Abstract This chapter examines the energy transition and climate policy landscape in the Galapagos Islands through the Social-Ecological-Technological Systems (SETs) framework. Despite national goals and international support, the islands remain heavily reliant on fossil fuels, particularly for transportation and electricity. Through policy analysis of six key documents, the chapter explores barriers to renewable energy adoption, including ecological constraints, social dynamics, and technological limitations. It identifies hybrid systems as the most promising path forward and offers guiding principles for sustainable development. The chapter emphasises the importance of integrated policymaking that balances ecological preservation, technological innovation, and community engagement in remote island contexts.

Introduction

This chapter delves into the complex interplay between the energy landscape, climate action, and the ecological and social integrity of the Galapagos Islands. As a remote ecosystem, the Galapagos faces distinctive challenges in balancing development, tourism, and environmental preservation. A critical issue facing the Galapagos is its

A. Doyon (✉)
Simon Fraser University, Burnaby, BC, Canada
e-mail: andreanne_doyon@sfu.ca

M. I. L. Plúa
Universidad San Francisco de Quito (USFQ), Quito, Ecuador

C. Cheng
Penn State University, University Park, PA, USA

Y. Ko
University of Michigan, Ann Arbor, MI, USA

J. Dunn Insua et al. (eds.), *Human Galapagos*, Social and Ecological Interactions in the Galapagos Islands, https://doi.org/10.1007/978-3-032-08800-0_6

overwhelming dependence on fossil fuels. 86% of the electricity in the islands is generated through diesel-burning processes (Galapagos Government Council 2021) This dependence on fossil fuels extends to terrestrial and maritime transportation, heavily used both by residents and the thriving commercial and tourism sector. In particular, a large portion of emissions in the Galapagos comes from transportation, which accounts for 77% of the total (Ministerio de Elecricidad y Energia Renovable 2017), with significant contributions from marine transportation, including water taxis that are integral to the tourism industry (Icaza-Alvarez et al. 2022) In response, several energy efficiency and zero-carbon initiatives have been introduced, such as the 'Galapagos Islands Zero Fossil Fuel Initiative'. In addition, there has been a variety of academic research conducted on modeling transitions to renewable energy for the islands (e.g., Arévalo et al. 2022; Icaza-Alvarez et al. 2022; Villacreses et al. 2023).

Despite the urgent need for a shift towards renewable energy sources, the Galapagos Islands have a surprisingly low penetration of renewable energy. Transitioning to renewable energy in the Galapagos requires addressing numerous challenges. Increased energy production from renewables could lead to land use conflicts (Ko et al. 2022), especially considering the archipelago's rich biodiversity and the fact that 97% of its land is designated as a national park. Technological limitations, such as low energy storage capacity and a grid system not fully adapted to renewable resources, compound these challenges. Additionally, there are social and policy barriers, such as energy costs equal to those of the mainland due to heavy subsidies, and local communities often prioritize immediate needs over environmental values. There are, however, emerging opportunities. For instance, the growing interest in eco-friendly tourism offers a potential revenue stream to support green initiatives, including the adoption of electric engines in transportation (Carvache-Franco et al. 2021). This chapter explores how renewable energy systems can be introduced and operated in the Galapagos, considering human and social perspectives through a review of six national and local policies and documents.

Sustainable development over decades has been undermined from systemic injustice and climate change impacts, which requires a systems approach to enhance community resilience, address climate justice, and achieve sustainability in economy, equity, and environment (Cheng 2022). Using the Social-Ecological-Technical Systems (SETs) framework (McPhearson et al. 2022) this chapter will use a policy and document analysis (Patton et al. 2015; Runhaar et al. 2006) to evaluate current energy and climate policy and to better understand the nuances of an energy transition in the Galapagos. This approach recognizes that energy and climate actions in the Galapagos are not just technological issues but require a more holistic, systematic approach. The Galapagos Islands are at a critical point where decisions made today will have lasting implications for both their ecological integrity and the well-being of their inhabitants. Understanding the energy transition in the context of this unique environment is not just a local necessity, but a matter of global significance. This chapter aims to provide insights and guidelines that will aid in navigating this complex and essential transition, particularly in remote islands. As well as offer guiding principles for future policy and action that harmonize the needs of this unique ecosystem with those of its inhabitants and the global community.

Energy and the Galapagos Islands

The Galapagos Islands' transition to renewable energy sources started as a way to mitigate environmental impacts and reduce their significant reliance on fossil fuels. The islands' transition is especially important as diesel-based electricity generation not only emits greenhouse gases but also increases the risk of diesel spillages. In 2001 this risk became a reality when the diesel tanker "Jessica" spilled approximately 75,000 gallons of fuel oil and 70,000 gallons of diesel along the coast of the Galapagos Islands (Global Sustainable Electricity Partnership 2013; Fundación Natura and World Wildlife Fund 2001). This spill, along with international pressure, led to the establishment of the Galapagos Islands Zero Fossil Fuel Initiative (CCFG in Spanish) in 2007, which included a complete elimination of fossil fuel use by 2020. Therefore, the Ecuadorian government commenced various solar and wind projects alongside additional initiatives, such as improving the efficiency of refrigerators, introducing electric vehicles (Consejo de Gobierno del Regimen Especial Galapagos 2016), and promoting induction ovens (Global Sustainable Electricity Partnership 2013). Nevertheless, these initiatives failed to achieve their intended objectives, as 83% of the energy grid remained dependent on fossil fuels (Insituto de Investigación Geológico y Energetico 2018).

As the 2020 deadline approached and the Galapagos Islands Zero Fossil Fuel Initiative failed to achieve its objectives, the Ecuadorian Government introduced the Galapagos Islands Strategic Plan for 2030. This plan advocates for sustainable development goals and introduces future projects aimed at achieving net-zero targets. Specifically, goal H32 promotes a transition towards a zero fossil fuel energy model and identifies 13 future projects aimed at increasing renewable energy production in San Cristóbal, Santa Cruz, Isabela, and Floreana (Galapagos Government Council 2021). However, the development of the proposed projects has been delayed due to the COVID-19 pandemic, and other unspecified factors. Thus far, from that plan only one photovoltaic (PV) plant has been implemented on San Cristobal Island (ARCERNNR), and five additional systems: two PV plants, two energy storage systems, and one dual-thermal plant are currently being planned for Santa Cruz and Isabela Islands. There is no further information or projected timelines available for the remaining seven projects (pers. communication, 2021). According to the plan, renewables in The Galapagos will have a nominal power of 41.83 MW, reaching 61% of the total energy grid in 2030. However, as of 2024, the Galapagos Islands has a nominal power use of 35.98 MW, with 23% from renewables and 76% coming from thermoelectric power plants (ARCERNNR 2022).

The region encompasses 11 renewable power plants distributed across five islands: San Cristobal, Santa Cruz, Isabela, Floreana, and Baltra. Of these, nine are photovoltaic systems, while the remaining two are wind systems. Most of these projects have been financed and implemented with the help of international alliances. For instance, Isabela Island features a hybrid-energy grid, which included an investment of approximately US$17 million financed by German Cooperation through the German Development Bank KFW (CGREG 2018). Similarly, San Cristóbal's hybrid-wind power system was sponsored by the Global Sustainable Electricity Partnership, an international alliance of utilities and electricity sector leaders (Global Sustainable Electricity Partnership 2016). Simultaneously, the private sector has also shown interest in the decarbonization of the islands. A prime example is the Conolophus Project, it is expected to be the biggest interconnected PV system in The Galapagos and will cost 63 million USD, funded entirely with private capital (MERNNR 2021).

Recently implemented projects have demonstrated that hybrid power systems offer the most promising results for meeting the energy demands in the Galapagos Islands. Given that renewable energy sources alone are currently insufficient to fully satisfy the islands' energy requirements, Hybrid Power Generation Systems (HPS) offer significant technical, economic, and environmental advantages over traditional energy systems (Llerena-Pizarro et al. 2019). Various analyses have evaluated different configurations of these systems and concluded that HPS could be suitable for isolated regions of interconnected electrical systems, such as the Galapagos archipelago (Roberts et al. 2018; Benalcazar et al. 2020). Current HPS in the islands are fossil fuel based, therefore several studies have proposed the following combinations: PV + Diesel (Garces-Palacios et al. 2024), PV + Wind (W) + Diesel (Arévalo et al. 2021; Cano et al. 2021), PV+ W+ Ecuadorian biodiesel, PV+ W+ Chinese biodiesel (Eras-Almeida et al. 2020), PV + W + Pumped Storage Hydro Systems (HP) (Cano et al. 2021) and PV + W + Biogas ((Llerena-Pizarro et al. 2019). From the studies, an 18 to 28.5 MWp system is believed to be the most promising, with the combination of solar/biogas power systems as an alternative to support the zero-fossil fuel initiative. It is also worth noting that alternative resources, such as geothermal energy and wave power, have received less attention and remain unexploited in the Galapagos Islands.

Galapagos Climate and Energy Policies

There are several different national and local policies and documents. In our policy and document analysis, six documents serve as the data for the SETs analysis, and the content of this chapter. They are: the National Energy Efficiency Plan 2016–2035; EmissionsGalapagos2019; Datos Balance Energético IIGE; Galapagos 2030; Energy Production Galapagos; and Energy Summary 2022. These documents were selected because they either provide strategic actions and goals to support an energy transition for the Galapagos Islands, or they provide data related to the islands' emissions, energy use, and energy production. Table 6.1 provides the name, author, and date of each document, as well as the energy and climate related content.

Table 6.1 Galapagos climate and energy policies and associated systems

Name of document	Author	Date	Climate and energy related content	SETs
National Energy Efficiency Plan 2016–2035	Ministerio de Electricidad y energía renovable (Ministry of Electricity and renewable energy), Inter-American Development Bank, Technical cooperation to support the preparation of the National Energy Efficiency Plan of Ecuador.	2017	National plan to implement energy efficiency projects, to reduce energy demand in the following years.	T
Emissions Galapagos 2019	Cristina Mateus, Daniela Flor, Christian A. Guerrero,· Ximena Córdova, Fátima L. Benitez, · René Parra,· Valeria Ochoa-Herrera	2019	Summarizes emissions (primary pollutants and GHGs) from all sectors present in the Galapagos for 2019	E
Balance Energético de la provincia de Galápagos (Galapagos Province Energetic balance)	Instituto de Investigación Geológico y Energético—IIGE (Geological and Energy Research Institute)	2018	Data for the Galapagos Islands regarding energy production, fuel consumption and emissions by sector.	E + T
Energy Production Galapagos	Ministerio de Energía y Recursos Naturales No Renovables (Ministry of Energy and Non-Renewable Natural Resources)		Percentages of energy production by type of power plant	T
			Historical production of renewable energy on the islands	E + T
			Diesel imports to Galapagos from 2001 to 2020	S
			Differences between islands in their installed and operational energy capacity as of 2021	T
Galapagos 2030	Galápagos Government Council; Technical Secretariat Planifica Ecuador; Ministry of the Environment and water; Ministry of Tourism, Ministry of Agriculture and Livestock	2021	Summary of energy related goals	S

(continued)

Table 6.1 (continued)

Name of document	Author	Date	Climate and energy related content	SETs
Energy Summary	Agencia de Regulación y Control de Energía y Recursos Naturales no Renovables—ARCONEL (Agency for the Regulation and Control of Energy and Non-Renewable Natural Resources)	2022	Energy production in the Galapagos in 2022 by each type of power plant	T
			Energy demand by sector in the Galapagos in 2022	S + T
			Electric Generation by source and location	E + T
			Electric Generation infrastructure information (i.e., cost, finance agreement, area, etc.)	S + T
			Current battery storage systems at the Galapagos	T
			Summary of studies on the expansion of electricity grid in Galapagos	S + T
			Future projects information	S
			Brief literature review that contextualizes the Galapagos energy system	S

Social-Ecological-Technological Systems (SETs) Framework

The complexity of sustainability on the Galapagos islands, which includes coping with climate change in concert with biodiversity conservation and habitat protection, as well as balancing sustainable priorities and people's livelihoods, is built upon the interdependence of Social-Ecological-Technological systems (SETs) (McPhearson et al. 2022). Any decision made about the islands will have consequences and impacts on people and the environment. The social, ecological, and technological systems are not in isolation. To address these challenges effectively, a holistic approach is imperative—one that recognizes the intricate interplay between social, ecological, and technological systems within the built environment. SETs is an integrated framework that acknowledges the dynamic interactions between human societies, natural ecosystems, and technological infrastructure within built environments. These systems operate within a complex web of dependencies and feedback loops, influencing one another in profound ways. Understanding and harnessing these interactions are essential for fostering sustainability and resilience in our built environments.

At the core of this framework lies the recognition of the inherent interconnectedness between human societies and the natural world. Social systems encompass the diverse array of actors, institutions, cultures, and behaviors that shape how communities interact with their environment. Ecological systems encompass the natural resources, biodiversity, and ecosystems upon which human societies depend for survival and well-being. Technological systems encompass the infrastructure, technologies, and innovations that facilitate human activities and shape the built environment.

Achieving sustainability and resilience within the built environment requires a multifaceted approach that considers the synergies and trade-offs between these three systems (Bush and Doyon 2019). Sustainability entails meeting the needs of the present without compromising the ability of future generations to meet their own needs (UN report 1984), while resilience involves the capacity to adapt and thrive in the face of change and uncertainty (Folke et al. 2016). Interactions between social, ecological, and technological systems can manifest in various ways within the built environment. For example, urbanization and industrialization may lead to habitat destruction, pollution, and resource depletion, impacting both human well-being and ecological health. Conversely, collocating solar PVs with pollinator plants can enhance biodiversity, mitigate urban heat island effects, and improve community well-being. Moreover, technological innovations such as renewable energy systems, smart grids, and green building technologies offer opportunities to reduce environmental footprints, enhance resource efficiency, and promote sustainable development. However, the successful implementation of such technologies relies heavily on social acceptance, governance structures, and institutional capacity (von Wirth et al. 2018).

The SETs framework is applied in this study to provide a more integrated and systems-oriented approach in order to understand the dynamics of the interactions between the systems in pursuit of sustainability and resilience in the built environment. When applying SETs to different contexts or scenarios, it helps to effectively manage the complexity of the interconnected systems by clarifying what is included or what are the parameters for each system. For this study, we are applying the SETs framework to the Galapagos energy transition. Below we include the parameters for each system, as well as for interaction systems (i.e., social + ecological). It is worth mentioning that there may be some overlap or repetition of components from one system to another.

- (S) Social system in energy transition: justice (recognition, procedural, and distributional), accessibility, behavior and practice change, education, economy, ownership, governance, management, diverse knowledges
- (E) Ecological system in energy transition: geography, topography, carbon emissions, air, water, biodiversity, ecosystem elements, structures, and functions
- (T) Technological system in energy transition: infrastructure systems, energy material innovations

- (S + E) Social + Ecological systems in energy transition: land use change and conflicts, limited developable land, use and conservation, natural resource management, ecosystem services, local knowledge of ecosystems
- (S + T) Social + Technological systems in energy transition: infrastructure planning and design, development, and management (proposals, plans, policies), knowledge of energy and technology, energy-related behaviors and practices, technological innovations
- (E + T) Ecological + Technological systems in energy transition: renewable energy (solar, wind, etc.), energy systems, impacts on biodiversity
- (S + E + T) Social-Ecological-Technological systems in energy transition: land use conflict, co-location of energy with other uses (e.g., urban settlement, agriculture), decentralized energy systems

SETs Analysis of Galapagos Climate and Energy Policies

The SETs framework is about the dynamic interactions between social, ecological, and technological systems, which has not been previously applied in the Galapagos Islands. Implementing such an analysis enables a more comprehensive evaluation of the broader impacts and feasibility of an energy transition within this unique context. This approach helps prevent potential adverse effects on the fragile Galapagos ecosystem and addresses the diverse perceptions of local communities, which are crucial for ensuring the sustainability of future energy solutions. Table 6.1 presents the results of the SETs analysis on the six selected policy documents. Figure 6.1 visualizes a SETs policy analysis of an energy transition for the Galapagos Islands.

A sociological analysis of the Galapagos Islands investigates how local perceptions and attitudes toward energy policies and consumption trends impact the feasibility of an energy transition. Current energy demand reveals the resistance to change and the degree of social acceptance of government-proposed policies. The Galapagos' reliance on diesel imports from the mainland makes their energy supply vulnerable to global disruptions. Additionally, the islands' heavy dependence on tourism complicates the potential for an energy transition, as any changes to the energy system could affect daily activities and livelihoods, especially commerce.

An ecological analysis assesses the environmental impacts of the current energy systems in the Galapagos Islands, focusing on their dependence on diesel imports. This reliance not only affects energy security, but also has significant ecological consequences, such as greenhouse gas emissions from both energy production and transportation. Current diesel consumption patterns and past experiences of importation delays have shown that the islands are heavily dependent on mainland supplies. Understanding these impacts is crucial for evaluating alternative energy technologies that could reduce harm to the natural environment and wildlife. When the environmental footprint of current systems is compared with potential renewable energy solutions, the potential for preserving the Galapagos' unique ecological heritage increases significantly with an energy transition.

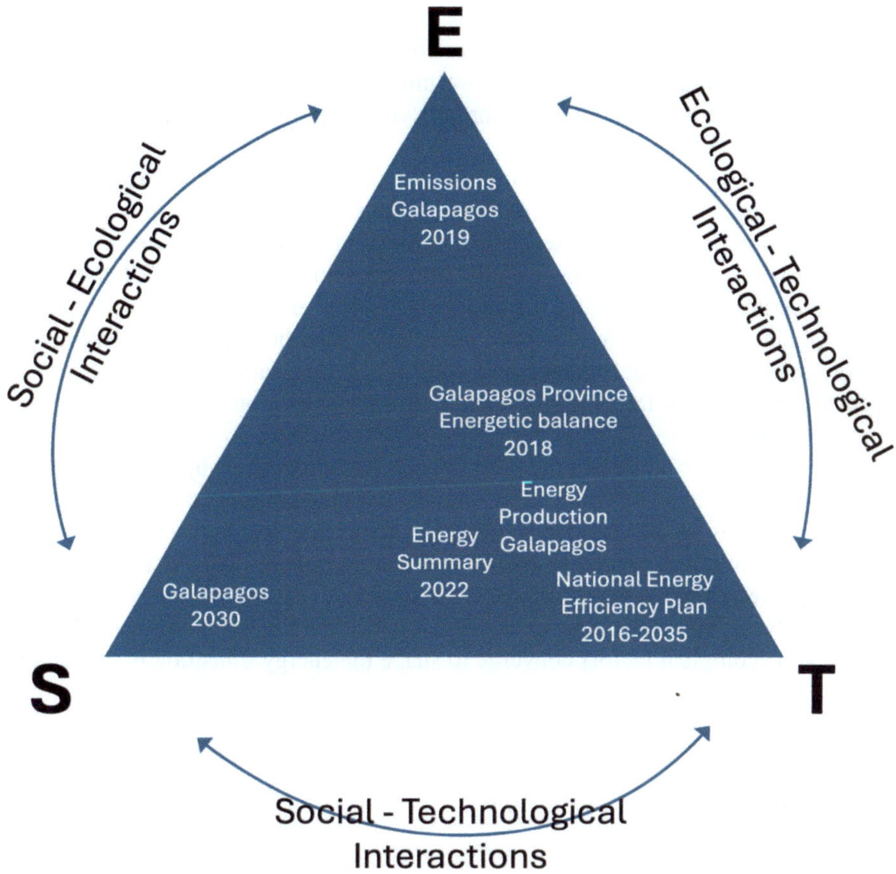

Fig. 6.1 A SETs policy analysis of an energy transition for the Galapagos Islands. (Modified from McPhearson et al. 2022)

The technological analysis assesses the current energy capacity and infrastructure of the Galapagos Islands, focusing on how effectively existing systems meet the islands' energy demands. The primary challenges in the Galapagos will not necessarily stem from the availability of global technologies, but rather from the practicalities of their implementation. These challenges include limited local expertise and the economic constraints characteristic of a small developing country. Furthermore, it is important to take into account the implications of introducing advanced and specialized technologies, considering both the need for external expertise and the potential social impacts of increasing the workforce involved in energy system development. Understanding these factors is crucial for developing a viable and sustainable energy transition strategy for the Galapagos Islands.

The interconnection of the analysis of the systems mentioned above reveals different relationships between aspects in the Galapagos Islands. For example, the interplay between technological advancements and social dynamics may necessitate additional labor, potentially raising concerns among locals about increased population and its implications for the islands' sustainability. This concern is further complicated by recent government efforts to manage tourism through higher entry fees, which creates a complex dynamic between local perceptions and broader energy sustainability goals.

Examining these interactions offers a better understanding of how technological innovations intersect with social realities. It is essential to involve the local community in participatory processes before implementing new energy measures. This approach will reveal disparities in equity and accessibility, particularly between the tourism-centric areas of Santa Cruz and San Cristóbal and the more remote agricultural highlands. Understanding how population distribution and perceptions of energy systems affect acceptance and support for renewable energy will provide valuable insights into the practical and cultural considerations necessary for a successful transition.

The analysis also addresses the social-ecological dimensions by considering how feelings of neglect and disconnection from the mainland influence local perceptions of future policies. In the Galapagos, where residents may feel isolated from national decision-making, it is crucial to ensure that energy policies address their concerns and are perceived as viable solutions. By exploring these complex interactions, the SETs analysis will contribute to a deeper understanding of how technological, social, and ecological factors converge to shape the energy transition in this unique context.

Discussion

The SETs framework provides a novel and integrative approach to understanding the energy transition challenges and opportunities in the Galapagos Islands. By examining the interplay between social, ecological, and technological systems, the SETs framework offers several new insights that are critical for achieving sustainable and resilient energy policies.

Holistic Understanding of Energy Transitions—The SETs framework moves beyond a purely technological or ecological perspective by incorporating social dimensions into the analysis. This comprehensive view highlights that energy transitions are not just about deploying new technologies, but also involve understanding and integrating human behaviors, social norms, governance structures, and policy frameworks. In the Galapagos, this means recognizing how local communities' needs, values, and economic activities intersect with environmental goals and technological capabilities, thereby offering a more nuanced approach to policy-making.

Interconnectedness of Systems—The framework emphasizes that social, ecological, and technological systems are deeply interconnected and influence each other in complex ways. For example, in the Galapagos, the push for renewable energy must consider land use conflicts (ecological concerns) and the existing energy-related behaviors of residents and tourists (social concerns). By acknowledging these interdependencies, the SETs framework allows for the identification of potential synergies and trade-offs. For instance, promoting eco-friendly tourism can simultaneously support renewable energy adoption (technological) and environmental conservation (ecological) while fostering social acceptance and economic benefits for local communities.

Integrated Policy Development—The SETs framework guides the development of integrated policies that address the multi-dimensional nature of energy transitions. In the case of the Galapagos, as shown in Fig. 6.1, each policy has its strength in managing one system yet can fall into siloed policies and procedures. Understanding the gaps of current policies in the interconnectedness in SETs, this approach calls for policies that balance the islands' biodiversity conservation efforts with the technological requirements for renewable energy systems and the social imperatives of community involvement and equitable access. This framework encourages policymakers to design adaptive, flexible strategies that can respond to the dynamic interactions between systems, ensuring that energy policies remain relevant and effective in the face of changing social, ecological, and technological conditions.

Addressing Systemic Barriers and Enhancing Resilience—By analyzing the barriers to renewable energy adoption through the SETs lens, new insights emerge about how to overcome systemic challenges. For example, understanding the role of governance structures (social system) and technological limitations (technological system) can help develop targeted interventions that improve institutional capacity and technological readiness. Additionally, the framework helps in building resilience by promoting a diversified approach to energy sources and solutions. This includes exploring hybrid systems that combine renewable sources with traditional energy forms, thereby reducing dependency on any single system and enhancing overall resilience against disruptions.

Facilitating Stakeholder Engagement and Social Acceptance—The SETs framework underscores the importance of involving local communities and stakeholders in the energy transition process. It provides insights into how social acceptance can be fostered through inclusive governance, transparent decision-making, and aligning energy initiatives with local priorities and values. In the Galapagos, leveraging the SETs framework means engaging communities in conversations about energy futures, ensuring that they are active participants in the transition rather than passive recipients of externally imposed solutions.

Limitations of This Study

The study relies on secondary data and lacks direct input from local communities, which restricts its understanding of social acceptance and engagement. This gap limits the effectiveness of applying the SETs framework in capturing the full range of local perspectives. This study should be viewed as an initial SETs assessment or proof of concept of the application of the SETs framework. Future research could build upon this study and include qualitative data from local stakeholders. The study does not thoroughly explore alternative energy sources like geothermal or wave energy, which may be well-suited to the Galapagos. This limitation affects the comprehensive application of the SETs framework, potentially missing viable options for sustainable energy transitions.

Conclusion

An energy transition in the Galapagos Islands will need to be a delicate balance between social equity, ecological preservation, and technological innovation. Achieving these goals will require ongoing commitment from various levels of government, and multiple stakeholders. Policies must be adaptable to evolving technological capabilities and social dynamics, fostering community engagement and ensuring equitable access to the benefits of renewable energy. The application of the SETs framework to the Galapagos Islands' energy transition offers new insights that are crucial for navigating the complex landscape of sustainable development. By recognizing the intertwined nature of social, ecological, and technological systems, the framework provides a more comprehensive and adaptive strategy for achieving a balanced and resilient energy future. This holistic approach not only addresses immediate energy needs but also lays the groundwork for sustainable growth that preserves the unique ecological and social fabric of the Galapagos.

References

Agencia de Regulación y Control de Energía y Recursos Naturales no Renovables (ARCERNNR) (2022) Panel de Infraestructura de Generación de Energía por Provincia (Energy Generation Infrastructure Panel by Province)

Arévalo P, Tostado-Véliz M, Jurado F (2021) Repowering feasibility study of a current hybrid renewable system. Case study, Galapagos Islands. Electricity 2(4):487–502. https://doi.org/10.3390/electricity2040029

Arévalo P, Cano A, Jurado F (2022) Mitigation of carbon footprint with 100% renewable energy system by 2050: the case of Galapagos islands. Energy 245:123247. https://doi.org/10.1016/j.energy.2022.123247

Benalcazar P, Suski A, Kamiński J (2020) Optimal sizing and scheduling of hybrid energy systems: the cases of Morona Santiago and the Galapagos Islands. Energies 13(15):3933. https://doi.org/10.3390/en13153933

Bush J, Doyon A (2019) Building urban resilience with nature-based solutions: how can urban planning contribute? Cities 29:102483. https://doi.org/10.1016/j.cities.2019.102483

Cano A, Arévalo P, Jurado F (2021) A comparison of sizing methods for a long-term renewable hybrid system. Case study: Galapagos Islands 2031. Sustain Energy Fuels 5(5):1548–1566. https://doi.org/10.1039/D1SE00078K

Carvache-Franco M, Carvache-Franco W, Manner-Baldeon F (2021) Market segmentation based on ecotourism motivations in marine protected areas and National Parks in the Galapagos Islands, Ecuador. J Coast Res 37(3):620–633. https://doi.org/10.2112/JCOASTRES-D-20-00076.1

Cheng C (2022) Intersection between systemic vulnerability, community resilience, climate justice and sustainability. In: Yang Y, Taufen A (eds) The Routledge handbook of sustainable cities and landscapes in the Pacific rim, pp 39–47. https://doi.org/10.4324/9781003033530-5

Consejo de Gobierno del Regimen Especial Galapagos (2016) Galápagos apuesta a la movilidad sustentable. Available at https://www.gobiernogalapagos.gob.ec/galapagos-apuesta-a-la-movilidad-sustentable/. Accessed: 10 Nov 2025

Consejo de Gobierno del Regimen Especial Galapagos (CGREG) (2018) En Isabela Avanza La Construcción Del Primer Proyecto Energético Híbrido Del Mundo. Available at https://www.gobiernogalapagos.gob.ec/en-isabelaavanza-la-construccion-del-primer-proyecto-energetico-hibrido-del-mundo/

Eras-Almeida A, Egido-Aguilera M, Blechinger P, Berendes S, Caamaño E, García-Alcalde E (2020) Decarbonizing the Galapagos Islands: techno-economic perspectives for the hybrid renewable mini-grid Baltra–Santa Cruz. Sustainability 12(6):2282. https://doi.org/10.3390/su12062282

Folke C, Biggs R, Norström AV, Reyers B, Rockström J (2016) Social-ecological resilience and biosphere-based sustainability science. Ecol Soc 21(3):41. https://doi.org/10.5751/ES-08748-210341

Fundación Natura, World Wildlife Fund (2001) The Jessica oil spill: God sleeps in Galapagos. In Fundación Natura, World Wildlife Fund, Galapagos Report 2000–2001, Quito

Galapagos Government Council (2021) Galapagos Islands Strategic Plan 2030. Available at: https://unidosporgalapagos.wordpress.com/2021/04/21/galapagos-island-strategic-plan-2030/. Accessed 2 Feb 2024

Garces-Palacios LE, Rodríguez-Gallegos CD, Vaca-Urbano F, Alvarez-Alvarado MS, Gandhi O, Rodríguez-Gallegos CA (2024) Optimal design of a hybrid solar–battery–diesel system: a case study of Galapagos Islands. Solar 4(2):232–245. https://doi.org/10.3390/solar4020011

Global Sustainable Electricity Partnership (2013) Galapagos San Cristóbal Island Wind Project 2007–2013

Global Sustainable Electricity Partnership (2016) Proyecto Eólico Isla San Cristóbal – Galápagos 2003–2016

Icaza-Alvarez D et al (2022) Decarbonization of the Galapagos Islands. Proposal to transform the energy system into 100% renewable by 2050. Renew Energy 189:199–220. https://doi.org/10.1016/j.renene.2022.03.008

Instituto de Investigación Geológico y Energético (2018) Balance energético de la Provincia de Galápagos 2018. Available at https://www.geoenergia.gob.ec/wpcontent/uploads/downloads/2020/05/balance_energetico_de_galapagos_2018.pdf. Accessed 10 Nov 2025

Ko Y, Coffman M, Mulvaney D, Copping A, Wang H, Barrett BFD (2022) Conflicts of greens' in renewable energy landscapes: case studies and a planning framework. In: Yang Y, Taufen A (eds) The Routledge handbook of sustainable cities and landscapes in the Pacific rim

Llerena-Pizarro OR, Micena RP, Tuna CE, Silveira JL (2019) Electricity sector in the Galapagos Islands: current status, renewable sources, and hybrid power generation system proposal. Renew Sust Energ Rev 108:65–75. https://doi.org/10.1016/j.rser.2019.03.043

McPhearson T, Cook EM, Berbés-Blázquez M, Cheng C, Grimm NB, Andersson E, Barbosa O, Chandler DG, Chang H, Chester MV, Childers DL, Elser SR, Frantzeskaki N, Grabowski Z, Groffman P, Hale RL, Iwaniec DM, Kabisch N, Kennedy C, Markolf SA, Matsler AM, McPhillips LE, Miller TR, Muñoz-Erickson TA, Rosi E, Troxler TG (2022) A social-ecological-technological systems framework for urban ecosystem services. One Earth 5(5):505–518. https://doi.org/10.1016/j.oneear.2022.04.007

Ministerio de Elecricidad & Energia Renovable (2017) Plan nacional de eficiencia energética 2016–2035 | ARIAE. Available at: https://www.ariae.org/servicio-documental/plan-nacional-de-eficiencia-energetica-2016-2035. Accessed 2 Feb 2024

Ministerio de Energía y Recursos Naturales no Renovables (MERNNR) (2021) PROYECTO: K020 MERNNR - Construcción de la planta fotovoltaica y la central térmica dual (aceite de piñón y diesel) Sistema Híbrido Isabela 2021. Available at: https://www.recursosyenergia.gob.ec/wp-content/uploads/2021/03/GPR-Hibrido-Isabela-2.pdf. Accessed 15 Jan 2023

Patton C, Sawicki D, Clark J (2015) Basic methods of policy analysis and planning. Routledge

Roberts JJ, Marotta Cassula A, Silveira JL, da Costa Bortoni E, Mendiburu AZ (2018) Robust multi-objective optimization of a renewable based hybrid power system. Appl Energy 223:52–68. https://doi.org/10.1016/j.apenergy.2018.04.032

Runhaar H, Dieperink C, Driessen P (2006) Policy analysis for sustainable development: the toolbox for the environmental social scientist. Int J Sustain High Educ 7(1):34–56

Villacreses G, Jijón D, Nicolalde JF, Martínez-Gómez J, Betancourt F (2023) Multicriteria decision analysis of suitable location for wind and photovoltaic power plants on the Galápagos Islands. Energies 16(1):29. https://doi.org/10.3390/en16010029

von Wirth T, Gislason L, Seidl R (2018) Distributed energy systems on a neighborhood scale: reviewing drivers of and barriers to social acceptance. Renew Sust Energ Rev 82(3):2618–2628. https://doi.org/10.1016/j.rser.2017.09.086

Chapter 7
Exploring the Interplay Between Endemic and Invasive Species in the Galapagos Islands: History of Colonization and Human Populations

Maria de Lourdes Torres, Martina Albuja-Quintana, Diego F. Cisneros-Heredia, Gabriela Pozo, María José Pozo, Pablo Alarcón-Bolaños, Mateo Dávila-Játiva, Emilia Peñaherrera-Romero, Daniel Die Morejón, Ariel Guerrero-Campoverde, Roberto J. León-E, and Daniel Velasco

Abstract Invasive species, introduced through centuries of human colonization, now threaten Galapagos ecosystems by outcompeting and displacing native species and reshaping the archipelago's unique biodiversity. Invasive plants often show greater adaptability and aggressiveness than their endemic counterparts or direct competitors, while non-native animals thrive in disturbed habitats, competing with and hybridizing with endemic fauna. Human activities have facilitated the introduction and spread of these species, irreversibly altering the archipelago's biodiversity. To protect this unique environment, urgent conservation efforts must involve local communities and implement effective tourism regulations to ensure the long-term preservation of the irreplaceable natural heritage of the Galapagos Islands.

Background

The current biodiversity composition of plant and animal species in the Galapagos Islands has been significantly shaped by anthropogenic activities dating back to the arrival of the first visitors to the islands. Over time, agriculture, trade, and tourism,

M. de Lourdes Torres (✉) · M. Albuja-Quintana · G. Pozo · M. J. Pozo · P. Alarcón-Bolaños
Laboratorio de Biotecnología Vegetal, Colegio de Ciencias Biológicas y Ambientales, Universidad San Francisco de Quito USFQ, Quito, Ecuador
e-mail: ltorres@usfq.edu.ec

D. F. Cisneros-Heredia · M. Dávila-Játiva · E. Peñaherrera-Romero · D. D. Morejón · A. Guerrero-Campoverde · R. J. León-E · D. Velasco
Laboratorio de Zoología Terrestre, Instituto de Biodiversidad Tropical IBIOTROP, Colegio de Ciencias Biológicas y Ambientales, Universidad San Francisco de Quito USFQ, Quito, Ecuador

J. Dunn Insua et al. (eds.), *Human Galapagos*, Social and Ecological Interactions in the Galapagos Islands, https://doi.org/10.1007/978-3-032-08800-0_7

among other activities, have been the gateway for the introduction of numerous species that now coexist alongside the native and endemic species of the archipelago.

Before human arrival, the unique biodiversity of the Galapagos originated from mainland America. Strong flyers like the ancestors of Galapagos finches likely reached the islands from the western coasts of South America (Lack 1983; Vincek et al. 1997). Powerful swimmers such as sea lions and penguins probably arrived aided by sea currents (Wolf et al. 2007; Kricher and Loughlin 2022). Many invertebrates, small vertebrates, and the ancestors of giant tortoises arrived on natural rafts of vegetation carried by sea currents (Caccone et al. 1999; Torres-Carvajal et al. 2014). Numerous plants likely reached the archipelago through wind dispersal, attaching to animal fur or feathers, and passing through digestive tracts (Gurevitch et al. 2006).

The Galapagos archipelago has a surprisingly recent history of interaction with non-native species, closely intertwined with human colonization (Mauchamp 1997; Causton et al. 2006; Tye 2006; González et al. 2008; Phillips et al. 2012; Keith et al. 2016; Cisneros-Heredia 2018). Since its discovery in 1535 and until the 1700s, the only visitors of the Galapagos were pirates and whalers who frequented the islands in search of fresh water, provisions, and shelter, probably introducing the first non-native species. It was common to release animals such as goats for hunting on the islands. Rats, likely stowaways on these ships, were among the first established introduced animals (Bassett 2009; Oxford et al. 2009; Toral-Granda et al. 2017; Bush et al. 2022).

During the nineteenth and early twentieth centuries, the first colonization attempts in the archipelago focused mainly on agricultural ventures (Astudillo 2018). Settlers introduced crops and modified the landscape to facilitate agriculture, inadvertently creating favorable habitats for other non-native organisms, such as invertebrates. This trend intensified in the mid-twentieth century as tourism flourished and crop production expanded to meet the rising demand. Consequent urban growth around key ports and tourism destinations, led to modified environments that provided fertile grounds for non-native plants and animals. The increase in population and trade of products in the archipelago facilitated the introduction of new species to the islands (Silberglied 1978; Toral-Granda et al. 2017; Sampedro et al. 2020).

The late twentieth century witnessed a growing recognition of the adverse effects of land modification and introduced species in the Galapagos. The Ecuadorian government's establishment of the Galapagos National Park in 1959 marked a milestone in conservation efforts, recognizing the uniqueness and delicate nature of the archipelago. In 1978, UNESCO declared the archipelago a World Heritage Site, further emphasizing its importance. In the twenty-first century, conservation efforts have intensified, focusing increasingly on the control of introduced species, the promotion of sustainable agricultural practices, and the conservation and restoration of Galapagos' unique ecosystems (Quiroga 2009; Kelley et al. 2019).

Invasive species, defined as introduced species that have successfully establish and disperse outside their native distribution (Diagne et al. 2021), usually take advantage of degraded and disturbed ecosystems to spread, colonize, and settle in the high number of available niches found in island ecosystems. These disturbed habitats are the result of human intervention like opened fields for agriculture and urban expansion towards habitable cities (Torres and Mena 2018). More than 1600 non-native species currently exist in the Galapagos Islands, established through intentional or unintentional introductions (Toral-Granda et al. 2017). This diverse group of invaders includes marine and terrestrial invertebrates, algae, terrestrial plants, and vertebrates (Causton et al. 2006; Tye 2006; Phillips et al. 2012; Cisneros-Heredia 2018; Carlton et al. 2019; Buchholz et al. 2020). Despite this growing body of knowledge, our understanding of invasive species ecology in the Galapagos remains partial. The complex interplay between different non-native species and the potential for future global change events introduce further uncertainties about their future dynamics. This lack of complete understanding presents a significant challenge for conservation efforts, as it hinders the development of effective strategies to control and eradicate these harmful invaders.

Plants, the most prevalent category of non-native species, exhibit a clear association with human settlements. It is now known that there are around 800 introduced plant species, 15% of which are considered invasive (Rivas-Torres et al. 2017). Introduced for food production, timber, or medicinal purposes, invasive plant species are mainly found in the highlands, where agriculture and anthropogenic activities have deteriorated the native vegetation (Mauchamp and Atkinson 2009). They now cover 2.2% of the Galapagos National Park and 28.5% of non-protected areas (Rivas-Torres et al. 2018; Laso et al. 2020). For example, the widespread presence of guava (*Psidium guajava*) and blackberry (*Rubus niveus*) is directly linked to agricultural efforts. These invasive plants aggressively compete with native flora, alter ecosystem processes, and can form dense thickets that limit available land for cultivation (Buddenhagen et al. 2004; Quinton et al. 2011; Rivas-Torres et al. 2017; Urquía et al. 2021). Ornamental plants, introduced for aesthetic purposes, show a distinct distribution pattern. They are most commonly found in green urban areas and private gardens within towns and settlements. Species like Lantana (*Lantana camara*), *Bougainvillea spectabilis*, Madagascar Periwinkle (*Catharanthus roseus*), and the Royal Ponciana (*Delonix regia*) have become established, adding splashes of color to human-modified landscapes but posing threats to native plant communities when they escape cultivation (McMullen 1999).

More than 600 invasive terrestrial invertebrates and 50 non-native vertebrates have been recorded in the Galapagos. About 85% of those invertebrates and 18 vertebrates have now become established into natural ecosystems of the archipelago (Toral-Granda et al. 2017; Cisneros-Heredia 2018; Ballesteros-Mejia et al. 2021). Urban areas are key points for non-native animals. Cats (*Felis catus*) and dogs (*Canis familiaris*), initially introduced as companions, have become feral predators, threatening native birds and reptiles. Even domesticated individuals can significantly threaten native fauna (Phillips et al. 2012; Cisneros-Heredia 2018). Non-native geckos have found ideal refuges in human infrastructure (Cisneros-Heredia 2018; Altamirano-Ponce et al. 2023). Beyond urban centers, agricultural activities aid in the spread of non-native animals. Feral livestock, like goats (*Capra hircus*) and pigs (*Sus scrofa*), free from natural predators, compete with native herbivores, leading to overgrazing and habitat degradation (Caccone et al. 1999). Invasive non-native animals have been documented in remote regions of the Galapagos, such as feral cats and black rats (*Rattus rattus*) that are a threat to the Critically Endangered pink iguana (*Conolophus marthae*) on the Wolf Volcano (Castaño et al. 2023). The impact is more alarming when invasive species transition from human-made environments into natural habitats. For example, the avian vampire fly (*Philornis downsi*) has devastated Galapagos bird populations, especially Darwin's finches (McNew and Clayton 2018).

In general, invasive species often undergo rapid genetic changes driven by selective pressures that allow them to persist, spread, and dominate their introduced environments (Sakai et al. 2001; Hierro et al. 2005). Human dispersal and anthropogenic activities have facilitated many introduction events of these invasive species (Torres and Mena 2018). Successful colonization and dispersal of these species is attributed to ecological (e.g., high dispersal rates) or environmental (e.g., low numbers of competitors) factors (Sakai et al. 2001), but recent studies suggest that genetic variability also plays a vital role in explaining effective introductions (Lee and Gelembiuk 2008), that can lead to significant negative impacts on biodiversity, ecosystem function, human welfare, and local economies. Poor public understanding of the damage and losses caused by invasive species worsens the problem, highlighting the urgent need to raise awareness before it is too late. This is essential to implement management strategies, which are often far less costly than the total damage caused by invasive species in ecosystems (Diagne et al. 2021).

Understanding the effects of invasive species on ecosystems requires a comprehensive, interdisciplinary approach. Molecular studies delve into the invader's genetic makeup, revealing details about population structure, potential hybridization with native species, and the origin of the invasion (Le Roux and Wieczorek 2009). This not only helps predict how the invader might spread and evolve but also underscores the importance of genetic diversity for invasion success (Miura 2007; Ward et al. 2008). Meanwhile, ecological studies offer insight into the invasion's impact on the ground. Through meticulous observations of behavior, distribution patterns, and competitive advantages, ecologists can elucidate how invasive species throw off the delicate balance of ecosystems, encompassing issues such as resource competition and disruption of pollination networks (Ward et al. 2008; Kenis et al. 2009). Molecular and ecological studies provide a better insight into invasive species, helping to develop strategies for effective control and mitigation, as well as contributing to the understanding of the dynamics of the effect of human settlements on native ecosystems and how anthropogenic activities modify the original landscape of the islands. By obtaining insights from both molecular and ecological studies, a comprehensive understanding of invasive species is possible, allowing the development of effective control and mitigation strategies and ultimately safeguarding global biodiversity.

For an engaging reading experience, this chapter is divided into two sections: invasive plants and invasive animals. The plant section focuses on two aspects: (1) comparing endemic and invasive species within the same genus, and (2) invasive

Table 7.1 Invasive plant species analyzed in this chapter

Invasive species	**Endemic counterpart**
Psidium guajava	*Psidium galapageium*
Solanum pimpinellifolium *Solanum lycopersicum*	*Solanum cheesmaniae* *Solanum galapagensis*
Invasive species	**Ecologically similar species**
Rubus niveus	*Miconia robinsoniana, Cyathea weatherbyana, Scalesia pedunculata, etc.*
Cedrela odorata	*Scalesia sp.*

species with no congeneric counterpart, yet still causing impacts on ecologically similar species (Table 7.1).

The animal section provides an overview of how various marine and terrestrial invasive species have been introduced and the impacts they have on endemic species. The primary focus is on how these introductions alter ecosystem dynamics.

Non-native Plant Species

Invasive Plants with an Endemic Counterpart in the Galapagos Islands

Interaction Between Guava (*Psidium Guajava*) and Guayabillo (*Psidium Galapageium*)

Guava (*Psidium guajava*) and guayabillo (*Psidium galapageium*) are closely related species within the Myrtaceae family and are both found on the Galapagos Islands. Guava is native to the tropical Americas and has been cultivated worldwide, while guayabillo is endemic to the Islands (Valdebenito 2018; Urquía et al. 2019). On inhabited islands, guava trees serve among others to delimit lands of different owners, provide shade, and as a source of food for various animals, helping to disperse seeds, while the guayabillo plays a crucial role in providing sustenance and shelter to native species, anchoring beneficial lichens, and providing nesting material for birds (Dal Forno et al. 2017; Valdebenito 2018; Urquía et al. 2019). Since these two related species cohabit certain areas of inhabited islands, it is interesting to analyze their interactions.

Research conducted on both species have revealed insights into their genetic diversity and population structure. A 2020 study that analyzed guayabillo populations from San Cristobal, Santa Cruz, and Isabela found moderate to high genetic diversity within the species and a clear population structure, indicating limited gene flow between islands due to oceanic barriers. Human presence is likely to increasingly influence future dynamics, highlighting the need for comprehensive conservation strategies (Urquía et al. 2020). In contrast, a study on the origin and genetic diversity of the guava in the Galapagos revealed relatively low genetic diversity compared to mainland counterparts, which is expected for invasive insular species, as well as a lack of population structure. Despite its low genetic diversity, multiple sources from the mainland to the invasive population's gene pool may have enhanced its adaptability and success in the archipelago. This study also proposes a colonization scenario for guava: this species would have been introduced initially to San Cristobal Island from a source population on mainland Ecuador. From San Cristobal it would have reached Floreana and then Isabela. Later, in another dispersal event, guava would have reached Santa Cruz from San Cristobal (Fig. 7.1) (Urquía et al. 2021).

Guava is one of the most aggressive invasive plants on the archipelago, and it poses a serious threat to the endemic guayabillo (Urquía et al. 2020; Torres et al. 2023). Introduced in the late 1800s, guava has aggressively spread across the highlands, displacing guayabillo (Urquía et al. 2019). Ongoing studies are currently investigating the impact of the interaction between guava and guayabillo. Preliminary findings suggest that guayabillo has been displaced by guava on San Cristobal Island, evidenced by its absence from areas dominated by guava. This displacement

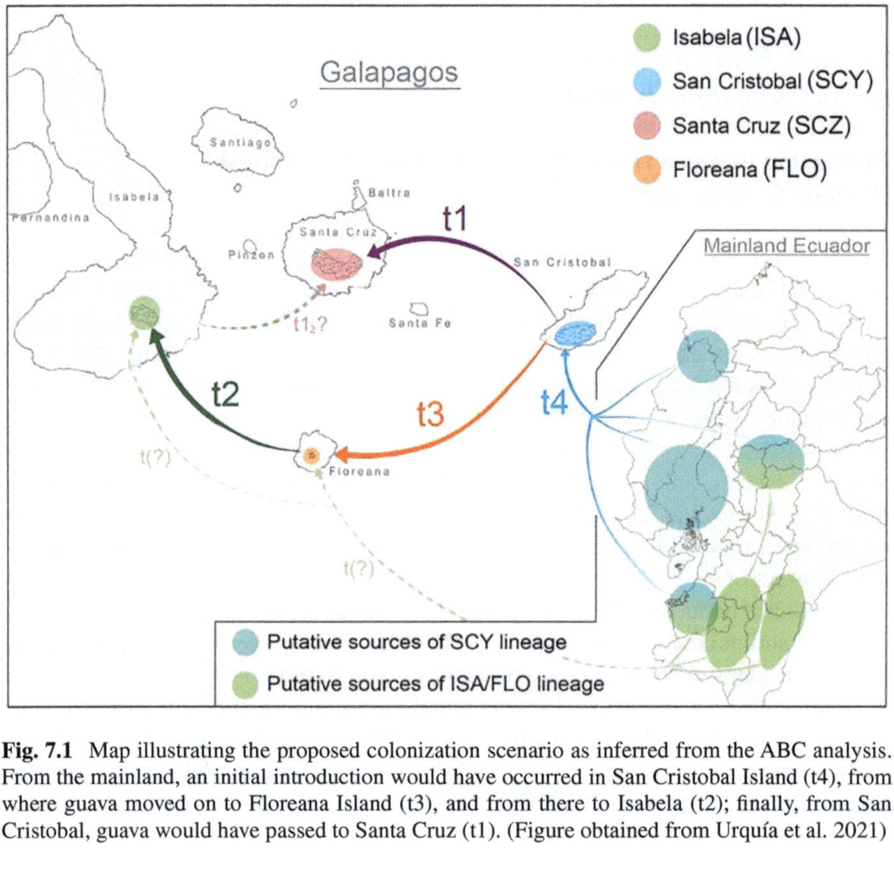

Fig. 7.1 Map illustrating the proposed colonization scenario as inferred from the ABC analysis. From the mainland, an initial introduction would have occurred in San Cristobal Island (t4), from where guava moved on to Floreana Island (t3), and from there to Isabela (t2); finally, from San Cristobal, guava would have passed to Santa Cruz (t1). (Figure obtained from Urquía et al. 2021)

phenomenon extends beyond San Cristobal, as islands with more recent guava introductions show a correlation between the arrival time of guava and the decline of guayabillo. Several factors may contribute to this decline: guava likely outcompetes guayabillo for resources, and it may disrupt guayabillo's reproduction through processes such as reproductive interference. This notion is supported by observations indicating that guayabillo populations near guavas have lower fruit and seed production. These findings emphasize the urgent need for conservation measures to mitigate the threat posed by guava to guayabillo and other endemic species in the Galapagos (Reatini et al. 2022).

The Interaction Between Introduced Tomatos and Endemic Tomatos

The Galapagos Islands are home to two endemic wild tomato species, *Solanum cheesmaniae* and *Solanum galapagensis*, which evolved in isolation on the islands and possess distinct genetic traits (Nuez et al. 2004). However, two introduced tomato species are also found today: *Solanum pimpinellifolium* and *Solanum lycopersicum*.

The rapid expansion of urbanization and human development activities in the Galapagos has emerged as a critical threat to these endemic tomato species. A pattern highlighted by Gibson et al. (2020) reveals that nearly all sites that were historically recorded populations of the endemic *S. cheesmaniae* and *S. galapagensis* could not be relocated during recent surveys. Additionally, this disappearance is within a relatively short timeframe. This decline isn't balanced by new populations emerging elsewhere, suggesting a systematic shift in species abundance. Notably, most missing endemic populations were near or within human-disturbed areas. Urbanization and habitat disruption are well-documented threats to Galapagos biodiversity. The disappearance of these locally adapted endemic populations from the islands represents an irreversible loss of valuable genetic diversity and unique trait variation important for crop improvement (Gibson et al. 2020; Torres et al. 2023).

Another study by Gibson et al. (2021) uncovered that the introduction of the invasive tomato species *S. pimpinellifolium* to the Galapagos Islands resulted in hybridization with the endemic congeneric species *S. cheesmaniae*. Genomic evidence reveals clear signals of recent hybridization and introgression between *S. pimpinellifolium* and *S. cheesmaniae*, including early-generation hybrids as well as older introgression events going back 4–12 generations. The genomic results point to several particularities of these hybridization processes between these two tomato species, but for the purposes of this chapter it is interesting to highlight the appearance of a new phenotype: the yellow color of the fruit in at least two invasive populations of *S. pimpinellifolium*, a color that is characteristic of the endemic species *S. cheesmaniae* (Fig. 7.2). The adaptive value of this new phenotype still needs to be analyzed to determine if the hybridization with the endemic species has facilitated the successful colonization of the invasive tomato in the archipelago. Exacerbating this issue, the same study found that increased human presence across the Galapagos archipelago is facilitating the proliferation of invasive *S.*

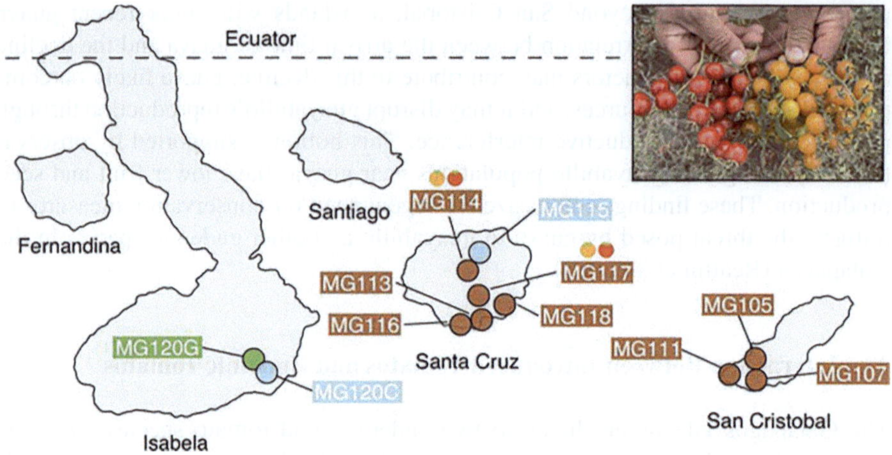

Fig. 7.2 Geographic distribution of focal sampling sites on the Galapagos Islands were polymorphic (red/orange) fruits representative of populations MG114 and MG117 of *S. pimpinellifolium* are found. (Figure obtained from Gibson et al. 2021)

pimpinellifolium and *S. lycopersicum* populations. Larger human population sizes are associated with higher abundances of these non-native tomatoes in surrounding disturbed environments ideal for their establishment and spread. The presence of invasive tomatoes not only intensifies resource competition but also enables further interspecific hybridization, introducing exotic genes that threaten the genetic integrity and survival of the endemic *Solanum* species (Gibson et al. 2021; Torres et al. 2023).

How the History of Invasion Relates to the Colonization of the Islands

Guava provides an interesting example of how the invasion history of a species is backed by the historical records of the colonization of the Islands. The arrival of humans in the archipelago and human activities such as trade, are directly linked to the spread of *P. guajava*. According to the genetic data, it is likely that the central Ecuadorian highlands are the primary source of the guava populations found on Isabela, Floreana, Santa Cruz, and San Cristobal (Urquía et al. 2021). This coincides with historical records that document that Galapagos received numerous immigrants from Tungurahua during early human settlement, namely indigenous people of the Salasaca culture (Grenier 2007). In addition, Manuel J. Cobos, one of the most famous settlers from the Galapagos, also came from this region of mainland Ecuador (Astudillo 2018). Historical records show that the first guava trees were planted on San Cristobal in the late nineteenth century by Cobos in his personal garden for ornamental purposes and that by the 1920s, Cobo's sugar plantation began to be displaced by large forests of *P. guajava* (Latorre 1997; Urquía et al. 2019). By the 1930's, guava was also documented in Floreana (Lundh 2006).

Although guava is poorly documented in Isabela records, this island appears to have been the next one to be invaded (Lundh 2006). Historical events suggest that guava may have been present in Isabela several decades before it was documented there in the 1950's. Antonio Gil, one of the first settlers of Isabela, arrived in 1897 after having lived in Floreana, and it is speculated that he might have carried guava seeds or fruit to Isabela (Latorre 1997; Urquía et al. 2019). Despite having witnessed the devastating effects of guava in San Cristobal, the few permanent settlers of Santa Cruz were unable to stop the guava invasion. Historical records show that by the 1950's, Santa Cruz had already been invaded by the guava (Lundh 2006). Together these findings suggest that humans, and not natural means, are responsible for the spread of guava between islands (Urquía et al. 2019).

The invasive wild tomato species *S. pimpinellifolium* introduction to the Galapagos Islands and subsequent invasion is also intimately tied to the history of human colonization and trade activities in the archipelago. Genomic evidence points to the primary invasive wild tomato population originating from the south-central region of mainland Ecuador, near Guayas and Los Rios provinces (Gibson et al. 2021). This aligns with records showing that the earliest major wave of human migration to the islands occurred after the 1949 Tungurahua earthquake, with colonizers arriving from central provinces like Los Rios (Toral-Granda et al. 2017). Moreover, most of the trade between Galapagos and the Ecuadorian mainland has centered around the major port city of Guayaquil (Guayas). Agricultural products from surrounding regions like Los Rios were likely inadvertent sources for the introduction of wild tomato seeds or plants to the islands through transportation of goods (Gibson et al. 2021). The recent demographic bottleneck and expansion signal detected in the *S. pimpinellifolium* genomic data corroborates a recent human-mediated colonization process. This confluence of genomic and historical evidence suggests human activities like migration and trade as critical vectors that facilitate the invasion of this alien tomato species across the Galapagos archipelago in recent times (Gibson et al. 2021).

Invasive Plants Without Endemic Counterparts in the Galapagos Islands (Fig. 7.3)

The Genetic Diversity of Blackberry (*Rubus niveus*) and Its Dispersal in the Islands

Rubus niveus (Thunb.) is a species that belongs to the Rosaceae family and is native to Asia (Quinton et al. 2011). The fruit of *R. niveus* is characterized by its pleasant flavor and rich nutritional content which are primary factors contributing to its global widespread consumption (Caidan et al. 2013). Since 1930, its presence has been reported in different areas of Oceania, Africa, and America (Centre for Agricultural Bioscience International 2019). Currently, *R. niveus* is also found on the inhabited islands of San Cristobal, Santa Cruz, Isabela, and Floreana, but also on

Fig. 7.3 Invasive plant species without endemic counterparts in the Galapagos Islands. (**a**) Individual of *Cedrela odorata* (Spanish Cedar) from Primicias Tortoise Reserve, Santa Cruz Island and (**b**) Individual of *Rubus niveus* (Blackberry) from San Joaquin, San Cristobal Island

the uninhabited island of Santiago (Quinton et al. 2011; Renteria et al. 2012). In the Galapagos Islands, it is considered a highly aggressive invasive species since its suggested introduction in the 1960s (Lawesson and Ortiz 1987; Gardener et al. 2013; Torres et al. 2023). Despite the absence of an endemic counterpart in the Galapagos Islands, *R. niveus* has been displacing native species such as *Miconia robinsoniana, Cyathea weatherbyana,* and *Scalesia pedunculata* by forming dense thickets and competing for space and resources (Quinton et al. 2011; Renteria et al. 2012; Global Invasive Species Database 2021).

The initial insights into the genetics of *R. niveus* were studied by Quinton et al. (2011) using two DNA non-coding regions as molecular markers. They provided the first phylogenetic tree grouping the Galapagos individuals with individuals from mainland Ecuador. Additionally, they found low nucleotide variability evidenced by sequence alignments, suggesting genetic proximity between the *R. niveus* populations analyzed. Alarcón et al. (2021) conducted a preliminary study of the genetic diversity of the blackberry in Galapagos populations using microsatellite markers, and found monomorphic alleles, suggesting low genetic differentiation within the populations examined. A recent study using genomic tools (Armijos et al. 2024) again reveals low levels of genetic differentiation among *R. niveus* populations on San Cristobal, Santa Cruz, and Isabela islands, as well as populations from mainland Ecuador. Other findings from this study suggest at least two separate introductions of blackberry from northern and central mainland Ecuador to the Galapagos Islands. The low genetic diversity found in all the mentioned studies surely reflects the reproductive biology of this species, probably with high rates of asexual propagation. Although future studies should delve deeper into this subject in order to use this information for the management of this invasive species, it is important to emphasize that sustained blackberry removal campaigns should be carried out in order to counteract the negative effects of this species both in agricultural and national park areas.

Cedrela odorata: A Case Study of Deliberate Plant Introduction That Outcompetes Nonrelated, Endemic Species

Cedrela odorata L. (Meliaceae) is the most renowned timber species of the genus *Cedrela*, which includes 18 species distributed across the Neotropical region (Pennington and Muellner 2010; Cavers et al. 2013). Valued for its aromatic, rot-resistant wood, *C. odorata* is threatened by illegal logging, leading to its inclusion in CITES Appendix II (2007) and classification as Vulnerable by the IUCN (2017).

Originally introduced to the Galapagos in the 1940s for its wood (Lundh 2006), *C. odorata* has now become the second most invasive tree species of the archipelago. It is found in the four inhabited islands of Galapagos, predominantly within agricultural areas (Laso et al. 2020). This invasive species has overtaken entire forests once dominated by endemic species (Rivas-Torres and Rivas 2018), resulting in decreased native plant diversity (Rivas-Torres et al. 2018). Unlike other species discussed in this chapter, *C. odorata* does not have a taxonomic counterpart with whom to compete or hybridize. Nevertheless, it directly competes with *Scalesia*, a canopy endemic tree unaccustomed to competition within its niche. As a result, *C. odorata* now dominates areas once covered by *Scalesia*, with the latter's population reduced to 1% of its original extent since the introduction of *C. odorata* (Rivas-Torres and Rivas 2018; Laso et al. 2020). The largest population (~1000 hectares) of *C. odorata* is found in the southwestern region of Santa Cruz (Rivas-Torres and Rivas 2018). Moreover, *C. odorata* not only outcompetes endemic tree species like *Scalesia* (Laso et al. 2020), but it also transforms Galapagos' ecosystems, alters community composition (Rivas-Torres et al. 2017), and propagates and colonizes new territories with its light-winged seeds (Mauchamp 1997; James et al. 1998). Its invasive success can be attributed to rapid germination (2–4 weeks), fast growth, and its ability to withstand extended dry seasons as a climate generalist (Cintron 1990). In some regions of its distribution, *C. odorata* sheds its leaves during the dry season to facilitate its survival in these conditions (Cavers et al. 2013).

Genetic diversity studies have revealed lower genetic diversity among *C. odorata* populations in Galapagos (He = 0.55) compared to mainland Ecuador (He = 0.81), but higher than other invasive island plant species (Urquía et al. 2019), suggesting multiple introductions events that could explain its dispersal and success as an invasive (Albuja-Quintana et al. 2024). As with other cases discussed in this chapter, human dispersal and colonization remain responsible for the ongoing effects of this introduced species.

How the History of Invasion Relates to the Colonization of the Islands

The studies described in this chapter about *Rubus niveus* and *Cedrela odorata* are interesting because they give us an idea of the role that genetic diversity, human colonization, and trade play in the dispersal and success of invasive species.

In the case of *R. niveus*, the available information suggests that despite the low levels of genetic diversity in *R. niveus* populations, they underwent a rapid range expansion since their arrival on the islands. The low genetic diversity found in populations of blackberry both in the mainland Ecuador and in the Galapagos Islands opens an interesting path to explore ways of managing this species, since it may be possible to envision forms of biological control that could be effective given this genetic characteristic. The relatively recent invasion of the blackberry in the Galapagos Islands reflects how human activities, including the willingness of settlers to replicate mainland lifestyles, can lead to impacts on island ecosystems.

According to the study by Albuja-Quintana et al. (2024), the agricultural areas of Santa Cruz seem to be one of the initial sites of introduction of the *C. odorata* in the Galapagos due to it being one of the main ports that received vessels from mainland Ecuador where seeds of *C. odorata* could have arrived (Lundh 2001). Nonetheless, San Cristobal also seems to be a point of entry for *C. odorata* since this island was also used as a port, due to its closer proximity to the mainland (Quiroga 2013). Being the most populated island during the mid-1900s (Lundh 2001), San Cristobal frequently received people, letters and goods from mainland Ecuador which could have included *C. odorata* seeds (Quiroga 2013). After these initial introductions, *C. odorata* appears to have arrived to Isabela and Floreana from either San Cristobal or Santa Cruz (Albuja-Quintana et al. 2024) probably as a result of the movement of settlers between islands.

Non-native Animal Species

Non-native Terrestrial Invertebrates

While vertebrates often dominate discussions on non-native species in the Galapagos, terrestrial invertebrates play a crucial, yet often overlooked role (Fig. 7.4). Spiders have been a focus of diversity and ecology research in the Galapagos since the early 1900s (Baert 2013; Buchholz et al. 2020). Surveys indicate that the archipelago is home to over 150 spider species, with half being endemic and at least 10% non-native (Baert and Maelfait 1986; Baert et al. 2008; Baert 2013, 2014; Buchholz et al. 2020). However, these figures may underestimate the actual number of non-native spiders. For instance, three non-native orb-weaver spider species of the family Araneidae (*Acacesia hamata*, *Argiope florida* and *Cyclosa triquetra*), previously unreported in the Galapagos, have been collected since 2019 in urban, periurban and agricultural areas of San Cristobal Island (Cisneros-Heredia and Reyes-Puig 2023). Additionally, some widespread orb-weaver spiders,

Fig. 7.4 Some terrestrial invertebrates found in the Galapagos. (**a**) *Cyclosa triquetra* (**b**) *Argiope argentata* (**c**) *Cheilomenes sexmaculata* (**d**) *Novius cardinalis*. (Photographs by (**a**) Ajott/iNaturalist (CC BY), (**b**) Mateo Dávila-Játiva, (**c**) David Torres and (**d**) Lucas Rubio (CC BY))

currently considered native to the Galapagos, such as *Argiope argentata* and *Neoscona oaxacensis*, might actually be allochthonous to the islands, introduced during the early periods of human habitation. Historically, these species were rare in mainland Ecuador, with few local records on the Pacific Coast. However, they have since significantly expanded their range along the Pacific coasts to the Andean highlands and Amazonian lowlands, indicating their geographic and ecological expansion potential (Levi 1968, 1993, 2004; Buchholz et al. 2020; Cisneros-Heredia and Reyes-Puig 2023). The possible impacts of non-native spiders on the native species and ecosystems of the Galapagos Islands remain uncertain. Yet, a recent study on pholcid spiders in Santa Cruz Island suggested a potential correlation, or even a causal link, between the spread of non-native spiders and the decline of native ones (Huber and Acurio 2022).

Extensive research exists on non-native ladybird beetles (Coccinellidae) worldwide due to their economic importance as agricultural biocontrol agents. However, Peck's (2006) landmark study on the Galapagos beetles reported three or four non-native ladybird species. Only one, the vedalia or cardinal ladybird beetle (*Novius cardinalis*), was identified at the species level. The remaining species were identified to genus (*Coccidophilus* sp. and *Hyperaspis* sp.) or subfamily (Hyperaspidinae). Ladybird beetles are well-suited for participatory science initiatives targeting non-native species (Cisneros-Heredia and Peñaherrera-Romero 2020; Venegas et al. 2021; Grez et al. 2022). Their distinct coloration makes them easily recognizable to the public, and their widespread use as biocontrol agents against garden and agricultural pests makes them one of the most familiar insect groups. Peñaherrera-Romero et al. (under review), combining field surveys and participatory science data, revealed that non-native ladybird beetles are prevalent in agricultural habitats in the highlands of San Cristobal Island, both in terms of species richness and abundance. In urban areas, non-native and native ladybird beetles coexist, while endemic species are increasingly rare and restricted to areas with native vegetation in the humid highlands. The two most common species recorded in the entire archipelago on iNaturalist, a participatory science platform by the California Academy of Science and National Geographic, are the native *Cycloneda sanguinea* and the non-native *Cheilomenes sexmaculata*, with most records coming from urban, periurban and agricultural areas in Santa Cruz, San Cristobal, Isabela, and Floreana islands. Records of other native or endemic species represent less than 10% of all observation uploaded to iNaturalist. These findings suggest a shift towards a predominance of non-native invertebrates in altered landscapes, possibly due to trophic cascade alterations (Greenslade et al. 2014; Convey and Lebouvier 2009; Convey et al. 2011).

In contrast to well-studied insect groups like ants, whose taxonomy and distribution are better known (e.g., Wauters et al. 2016), determining whether many other terrestrial invertebrate species, such as spiders and ladybird beetles, are native or introduced to the Galapagos islands is challenging due to the limited taxonomic work and historical or current data for many islands (Buchholz et al. 2020; Cisneros-Heredia and Reyes-Puig 2023). This raises concerns about the true status of several populations and highlights the need to determine the islands' native community composition promptly before it becomes homogenized.

How the History of Invasion Relates to the Colonization of the Islands

The exact arrival pathway of many non-native spiders and ladybird beetles in the Galapagos remains unclear (Peck 2006; Buchholz et al. 2020; Cisneros-Heredia and Reyes-Puig 2023; The excellent dispersal abilities of spiders complicate efforts to pinpoint their arrival pathway or even determine if they are native (Buchholz et al. 2020).

It is likely that most non-native spider and ladybird beetle species arrived as stowaways through one or various pathways, such as on boats or planes, on potted plants, or in fruit shipments, similar to reported cases elsewhere (Peck 2006; Causton et al. 2006; Toral-Granda et al. 2017; Cisneros-Heredia 2018; Buchholz et al. 2020; Cisneros-Heredia and Reyes-Puig 2023). Their arrival probably started as early as the first humans visited the islands; however, it most probably peaked in the twentieth century with the increment of urban and agricultural land use. Tourism may have played an important role since transport on boats can apparently carry species long distances from their original ranges, as seen with some spiders native to North America that arrived to the Galapagos on tourism ships departing from Florida (Cisneros-Heredia 2018).

One exception to inadvertent introductions of non-native species is the *Vedalia* ladybird beetle. This species was intentionally introduced to the Galapagos between 2002 and 2005 as a biocontrol agent for the cottony cushion scale (*Icerya purchasi*). It successfully fulfilled this role, leading to a 60–98% reduction in the scale population on the islands (Calderón-Alvarez et al. 2012; Hoddle et al. 2013). Nowadays, this species is widespread and present across human-made and natural ecosystems.

Non-native Reptiles

Some of the most renowned and best-studied inhabitants of the Galapagos Islands are reptiles, including iconic and unique species like the giant tortoises (*Chelonoidis* spp.) that give their name to the archipelago, marine iguanas (*Amblyrhynchus cristatus*), the only sea-faring lizards globally, and land iguanas (*Conolophus* spp.). However, this reptilian assemblage extends beyond these giants to include lesser-known groups such as Leaf-toed geckos (*Phyllodactylus* spp.), lava lizards (*Microlophus* spp.) and racer snakes (*Pseudalsophis* spp.) (Benavides et al. 2009; Torres-Carvajal et al. 2016; Rowe et al. 2019; Cisneros-Heredia and Reyes-Puig 2023). Seventeen species of geckos, with 13 endemic and four non-natives, are currently known from the islands. Most islands in the archipelago host one endemic species and up to four non-natives, except for San Cristobal and Isabela islands, which have two endemic species each (Torres-Carvajal et al. 2014; Altamirano-Ponce et al. 2023).

Non-native geckos likely arrived in the Galapagos islands via vessels mainly from South America, though they have widely different biogeographic origins. The dwarf gecko (*Gonatodes caudiscutatus*) is native to the Pacific lowlands of Ecuador and northern Peru (Sturaro and Avila-Pires 2013). Currently, it is established only in San Cristobal Island, occurring in small numbers in urban green areas in Puerto Baquerizo Moreno town and higher densities in agricultural areas in the highlands, particularly in El Progreso, where it has also established populations in natural areas (Hoogmoed 1989; Olmedo and Cayot 1994; Cisneros-Heredia 2018; Altamirano-Ponce et al. 2023). A recent study analyzing genetic diversity using a mitochondrial gene (16S) in populations from the Galapagos Islands and mainland Ecuador revealed low nucleotide diversity and low haplotype diversity in Galapagos populations, with just two haplotypes found in San Cristobal Island out of nine (Altamirano-Ponce et al. 2023). This contrasts with the coastal leaf-toed gecko (*Phyllodactylus reissii*), which is also native to the Pacific lowlands of western Ecuador and northwestern Peru but remains fairly uncommon in specific localities in urban and natural areas of the dry lowlands of San Cristobal, Santa Cruz, and Isabela islands (Hoogmoed 1989; Olmedo and Cayot 1994; Torres-Carvajal et al. 2014; Cisneros-Heredia 2018).

The common house gecko (*Hemidactylus frenatus*) and the mourning gecko (*Lepidodactylus lugubris*) (Fig. 7.5) originated much further away, in Asia and Oceania, respectively. They have had detrimental impacts on endemic geckos in other island ecosystems due to competition for space and food, disease transmission and predation (Cole et al. 2005; Callaway et al. 2011; Hoskin 2011; Swanepoel et al. 2022). Both species quickly established themselves and spread in large numbers across all populated areas of the Galapagos, replacing the previously common endemic geckos (Torres-Carvajal and Tapia 2011; Cisneros-Heredia 2018). The establishment and spread of non-native geckos in the archipelago have been facilitated by human-made structures that provide substrates, allowing invasive geckos to move faster and more easily than endemic ones. Dense populations of introduced geckos in touristic areas adjacent to natural ecosystems indicate their invasion potential, highlighting the urgent need to address conservation challenges facing endemic geckos and to develop strategies to mitigate and control impacts by non-native geckos (Cisneros-Heredia 2018; Oleas-Paz and Cisneros-Heredia 2019; Ramos-Rojas and Cisneros-Heredia 2019).

Fig. 7.5 Non-native geckos in the Galapagos archipelago. (**a**) *Hemidactylus frenatus* (**b**) *Lepidodactylus lugubris* (**c**) *Gonatodes caudiscutatus* (**d**) *Gonatodes caudiscutatus*. (Photographs by (**a**) Daniel Velasco and (**b–d**) Mateo Dávila-Játiva)

How the History of Invasion Relates to the Colonization of the Islands

The introduction pathway of all non-native geckos is likely as stowaways in cargo shipments (Hoogmoed 1989; Olmedo and Cayot 1994; Cisneros-Heredia 2018). The dwarf gecko was the first non-native gecko to arrive in the Galapagos, with records dating back to the late nineteenth century, when the first human settlements were established in San Cristobal Island (Garman 1892; Wright 1983; Hoogmoed 1989; Olmedo and Cayot 1994; Cisneros-Heredia 2018; Altamirano-Ponce et al. 2023). The coastal leaf-toed gecko at Santa Cruz Island arrived in the mid-1970s, probably on boats from coastal Ecuador. Upon their arrival, they became established and, for a long time, largely dependent on urban and agro-urban areas in populated islands of the Galapagos. In contrast, the mourning and common house geckos are more recent arrivals; the former arrived around the 1980s and the latter around the mid-2010s. Both are restricted to human-made ecosystems, including urban, touristic and some rural areas of San Cristobal, Santa Cruz, Isabela, and Floreana (Torres-Carvajal and Tapia 2011; Cisneros-Heredia 2018; Oleas-Paz and Cisneros-Heredia 2019; Ramos-Rojas and Cisneros-Heredia 2019).

Non-native Marine Species

The impacts of urban settlements and species introductions in the Galapagos extend beyond the terrestrial ecosystems to encompass the less studied marine environments. The islands' marine life thrives due to the convergence of warm equatorial and cold nutrient-rich currents, with upwelling processes further enhancing biodiversity (Witman and Smith 2003). The dynamic interplay of sea temperature seasonality and geographical variations, influenced by El Niño oceanographic anomalies, exert a profound influence on marine ecosystems (Edgar et al. 2004; Brandt et al. 2022; Rhoades et al. 2023). Despite their resilience to drastic oceanic variations, marine communities are vulnerable due to ongoing exposure to maritime connectivity. The introduction of over 50 non-native marine species, including algae and invertebrates, pose a significant threat (Carlton et al. 2019). These non-native species compete for essential resources such as rocky surface cover and food, disrupting native ecosystems (Schaffelke et al. 2006). Green grape algae (*Caulerpa* spp.) are a notable example of ecological disruption and rapid proliferation, outcompeting native algal and coral species (Williams and Grosholz 2002; Klein and Verlaque 2008; Keith et al. 2022). Other introduced organisms, like brown algae, cnidarians, polychaete worms, bivalves, bryozoans, and echinoderms, also pose potential threats due to their overgrowth and competitive interactions and predatory behaviors on local subtidal habitats, ports, protected bays and mangroves (Keith et al. 2016; Carlton et al. 2019). Similar to terrestrial ecosystems, native and alien species interact, altering the ecological structure of marine environments (Keith et al. 2016; Glynn et al. 2018; Riegl et al. 2019).

How the History of Invasion Relates to the Colonization of the Islands

Non-native marine species often arrived facilitated by marine traffic hotspots like ports of discharge and tourist destinations (Letschert et al. 2021). It is crucial to evaluate human activities' impacts on the establishment of non-native species in marine environments in the Galapagos. Stringent biosecurity measures and ongoing monitoring are necessary within the Galapagos Marine Reserve and the Galapagos National Park to prevent the introduction and spread of marine non-native species and preserve the unparalleled marine biodiversity of the archipelago (Toral-Granda et al. 2017; Carlton et al. 2019).

Conclusions

The legacy of the colonization of the Galapagos Islands is complex. It has led to the establishment of permanent human settlements and significant economic activity that has irrevocably altered ecosystems. Anthropogenic activities, persisting from the past until the present, pose a threat to the welfare of the wildlife inhabiting the Galapagos Islands. Invasive plant and animal species are now a major threat to native flora and fauna. The prevalence of invasive animal and plant species in urban areas, agricultural fields, and other human-altered environments can be attributed to several factors. Human structures often provide unintentional refuges, protecting these exotic species from harsh island conditions, like invasive geckos, which usually hide in human houses from environmental conditions like rain (Torres-Carvajal et al. 2016). These alterations also create ideal breeding grounds, as human-caused disturbances pave the way for the establishment and spread of non- native species in native ecosystems. Furthermore, compared to their native counterparts, these introduced species often exhibit a remarkable ability to adapt to these human-disturbed ecosystems. Species like *Psidium guajava*, for example, are used to and grow around human settlements both in the mainland and islands, while *Psidium galapagensis* is used to grow in undisturbed habitats (Torres-Carvajal et al. 2016).

The impact of these non-native species transcends individual populations. By disrupting food webs and altering habitat structure, they unravel the complex ecological tapestry of the Galapagos. This not only jeopardizes the survival of endemic species but also diminishes the ecological resilience of the islands as a whole. Urgent conservation efforts are needed to mitigate the threat posed by these invaders to ensure the continued existence of this irreplaceable natural wonder.

That said, it is also important to find the right balance between conservation and the well-being of the human populations inhabiting the Galapagos. Understanding their needs, offering education and healthcare will surely help to have an informed and resourceful population that is aware of the special place they inhabit and the urgent need for its preservation. Sharing research on invasive species with the local community could help raise awareness and support the development of management policies, similar to those implemented in other archipelagos. For example, in the French Polynesian Islands, surveys were conducted among people from diverse backgrounds to assess how feasible it would be to involve locals in prevention, monitoring, and control efforts targeting invasive species (Meyer et al. 2019). Approaches such as this could be replicated in the Galapagos Islands.

If we analyze the examples of invasive species described in this chapter, we can find some common ground: most of them are closely related to the migration of people from the mainland to the islands for various reasons throughout the history of colonization of the archipelago. Migrants transport items to their new home which give them comfort, whether this be domestic animals, ornamental, or edible plants. However, this can entail serious consequences, such as those mentioned in previous sections. In this way, as urban centers are created, new elements intermingle in the island landscape and in some cases spread aggressively within these urban areas, invade agricultural areas, and even escape to uninhabited locations that are part of the Galapagos National Park's protected areas.

The world is now irreversibly global, where tourism has become a giant source of income for many countries. This moment can be seen as a new form of colonization, not through conquest, but consumption. We occupy space, take photos, live experiences, and move on. Yet even this brief presence leaves a deep ecological footprint. Ironically, tourism depends on the very ecosystems it threatens. The more the people come to witness natural wonders like the Galapagos Islands, the more those ecosystems are in danger. Unless we recognize and limit this new form of colonization and actively work to reduce our ecological footprint, places like the Galapagos may not endure long.

References

Alarcón P, Torres ML, Pozo G, García MP, Ortega M, Rojas E et al (2021) Evaluación de marcadores microsatélites (SSRs) heterólogos en *Rubus niveus* para estudios de diversidad genética en las Islas Galápagos. ACI Avances en Ciencias e Ingenierías 13(2):20–20

Albuja-Quintana M, Rivas-Torres G, Rojas López KE, Asadobay P, Palacios Cuenca W, Vinueza G, Torres MDL (2024) Preliminary insights of the genetic diversity and invasion pathways of *Cedrela odorata* in the Galapagos Islands, Ecuador. Ecol Evol 14(7):e11723

Altamirano-Ponce L, Dávila-Játiva M, Pozo G, Pozo MJ, Terán-Velástegui M, Cadena CD et al (2023) First genetic insights of *Gonatodes caudiscutatus* (Reptilia, Gekkota) in the Galapagos Islands and mainland Ecuador. Biodiver Data J:11

Armijos CE, Cerca J, Alarcón P, Torres ML (2024) Population Genomics of the Invasive Blackberry (*Rubus niveus*) in the Galapagos Islands (Under Review)

Astudillo FJ (2018) Environmental and historical archaeology of the Galápagos islands: Archaeobotany of hacienda El Progreso, 1870–1920. Veg Hist Archaeobotany 27(5):737–751. https://doi.org/10.1007/s00334-018-0668-9

Baert L (2013) Summary of our present knowledge of the spider communities of the Galápagos archipelago. First analysis of the spider communities of the islands Santa Cruz and Isabela. Belg J Zool 14:159–185

Baert L (2014) Three new Eustala (Araneae, Araneidae) species from the Galápagos Islands (Ecuador). Eur J Taxon 86:1–18

Baert LL, Maelfait J-P (1986) A contribution to the knowledge of the spider Fauna of Galápagos (Ecuador). Bull Inst R Sci Nat Belg Entomol 56:93–123

Baert LL, Malfait J-P, Hendrickx F, Desender K (2008) Distribution and habitat preference of the spiders (Araneae) of Galápagos. Bull Inst R Sci Nat Belg Entomol 78:39–111

Ballesteros-Mejia L, Angulo E, Diagne C, Cooke B, Nuñez MA, Courchamp F (2021) Economic costs of biological invasions in Ecuador: the importance of the Galapagos Islands. NeoBiota 67:375–400

Bassett CA (2009) Galápagos at the crossroads: pirates, biologists, tourists, and creationists battle for Darwin's cradle of evolution. National Geographic Books

Benavides E, Baum R, Snell HM, Snell HL, Sites JW Jr (2009) Island biogeography of Galápagos lava lizards (Tropiduridae: *Microlophus*): species diversity and colonization of the archipelago. Evolution 63(6):1606–1626

Brandt M, Silva-Romero I, Fernández-Garnica D, Agudo-Adriani E, Bove CB, Bruno JF (2022) Top-down and bottom-up control in the Galápagos upwelling system. Front Mar Sci 9:845635. https://doi.org/10.3389/fmars.2022.845635

Buchholz S, Baert L, Rodríguez J, Causton E (2020) Spiders in Galapagos – diversity, biogeography and origin. Biol J Linn Soc 130(1):41–48. https://doi.org/10.1093/biolinnean/blaa019

Buddenhagen CE, Renteria JL, Gardener M, Wilkinson SR, Soria M, Yánez P et al (2004) The control of a highly invasive tree *Cinchona pubescens* in Galapagos. Weed Technol 18(sp1):1194–1202

Bush MB, Conrad S, Restrepo A, Thompson DM, Lofverstrom M, Conroy JL (2022) Human-induced ecological cascades: extinction, restoration, and rewilding in the Galápagos highlands. Proc Natl Acad Sci 119(24):e2203752119

Caccone A, Gibbs JP, Ketmaier V, Suatoni E, Powell JR (1999) Origin and evolutionary relationships of giant Galápagos tortoises. Proc Natl Acad Sci 96(23):13223–13228

Caidan R, Cairang L, Yourui S (2013) Simultaneous analysis of fatty acids in *Rubus niveus* Thunb. Fruits by HPLCMS/MS. Asian J Chem 25(4):1866–1870. https://doi.org/10.14233/ajchem.2013.13204

Calderón-Alvarez C, Causton CE, Hoddle MS, Hoddle CD, Van Driesche RG, Stanek EJIII (2012) Monitoring the effects of *Rodolia cardinalis* on *Icerya purchasi* populations in the Galápagos. BioControl 57:167–179

Callaway Z, Thomas A, Melrose W, Buttner P, Speare R (2011) *Salmonella* Virchow and *salmonella* Weltevreden in a random survey of the Asian house gecko, *Hemidactylus frenatus*, in houses in northern Australia. Vector-Borne Zoonotic Dis 11(6):621–625. https://doi.org/10.1089/vbz.2010.0015

Carlton J, Keith I, Ruiz G (2019) Assessing marine bioinvasions in the Galápagos Islands: implications for conservation biology and marine protected areas. Aquat Invasions 14(1):1–20. https://doi.org/10.3391/ai.2019.14.1.01

Castaño PA, Hanson CC, Campbell KJ, Carrión V, Fisher P, Ruell E et al (2023) Invasive rodent eradication on islands: assessment and mitigation of human exposure to rodenticides. Biol Invasions 25(3):653–671

Causton CE, Peck SB, Sinclair BJ, Roque-Albelo L, Hodgson CJ, Landry B (2006) Alien insects: threats and implications for conservation of Galápagos Islands. Ann Entomol Soc Am 99(1):121–143

Cavers S, Telford A, Cruz FA, Castañeda AJP, Valencia R, Navarro C, Buonamici A, Lowe AJ, Vendramin GG (2013) Cryptic species and phylogeographical structure in the tree *Cedrela odorata* L. throughout the Neotropics. J Biogeogr 40(4):732–746. https://doi.org/10.1111/jbi.12086

Centre for Agricultural Bioscience International (2019) *Rubus niveus* (Mysore raspberry). Retrieved from https://web.archive.org/web/20210506091150/https://www.cabi.org/isc/datasheet/107939

Cintron B (1990) *Cedrela odorata* L. Cedro hembra, Spanish cedar. In: Burns R, Honkala B (eds) Silvics of North America. U.S. Department of Agriculture, Washington, pp 250–257

Cisneros-Heredia DF (2018) The hitchhiker wave: non-native small terrestrial vertebrates in the Galapagos. In: Torres MDL, Mena CF (eds) Understanding invasive species in the Galapagos Islands. Springer International Publishing, pp 95–139. https://doi.org/10.1007/978-3-319-67177-2_7

Cisneros-Heredia DF, Peñaherrera-Romero E (2020) Invasion history of *Harmonia axyridis* (Pallas, 1773) (Coleoptera: Coccinellidae) in Ecuador. PeerJ 8:e10461

Cisneros-Heredia DF, Reyes-Puig C (2023) Natural history and conservation of the Galapagos Snake radiation. Islands Snakes Diver Conserv:157

CITES (2007) Examen de las propuestas de enmienda a los Apéndices I y II. Convención sobre el comercio internacional de especies amenazadas de fauna y flora silvestres. https://cites.org/sites/default/files/esp/cop/14/prop/S14-P33.pdf. Accessed 20 Nov 2022

Cole NC, Jones CG, Harris S (2005) The need for enemy-free space: the impact of an invasive gecko on Island endemics. Biol Conserv 125(4):467–474. https://doi.org/10.1016/j.biocon.2005.04.017

Convey P, Lebouvier M (2009) Environmental change and human impacts on terrestrial ecosystems of the sub-Antarctic islands between their discovery and the mid-twentieth century. In Papers and proceedings of the royal society of Tasmania (vol 143, no 1, pp 33–44)

Convey P, Key RS, Key RJD, Belchier M, Waller CL (2011) Recent range expansions in non-native predatory beetles on sub-Antarctic South Georgia. Polar Biol 34:597–602

Dal Forno M, Bungartz F, Yánez-Ayabaca A, Lücking R, Lawrey JD (2017) High levels of endemism among Galapagos basidiolichens. Fungal Divers 85(1):45–73. https://doi.org/10.1007/s13225-017-0380-6

Diagne C, Leroy B, Vaissière AC, Gozlan RE, Roiz D, Jarić I et al (2021) High and rising economic costs of biological invasions worldwide. Nature 592(7855):571–576

Edgar GJ, Banks S, Fariña JM, Calvopiña M, Martínez C (2004) Regional biogeography of shallow reef fish and macro-invertebrate communities in the Galapagos archipelago. J Biogeogr 31(7):1107–1124. https://doi.org/10.1111/j.1365-2699.2004.01055.x

Gardener MR, Trueman M, Buddenhagen C, Heleno R, Jäger H, Atkinson R, Tye A (2013) A pragmatic approach to the management of plant invasions in Galapagos. In: Plant invasions in protected areas: patterns, problems and challenges, pp 349–374

Garman S (1892) The reptiles of the Galapagos Islands. Bull Essex Institute 24:73–87

Gibson MJS, de Lourdes Torres M, Moyle LC (2020) Local extirpation is pervasive among historical populations of Galápagos endemic tomatoes. Evol Ecol 34(2):289–307. https://doi.org/10.1007/s10682-020-10035-3

Gibson MJ et al (2021) Introgression shapes fruit color convergence in invasive Galápagos tomato. elife 10:e64165. https://doi.org/10.7554/eLife.64165

Global Invasive Species Database (2021) Species profile: *Rubus niveus*. Retrieved de http://www.iucngisd.org/gisd/species.php?sc=1232

Glynn PW, Feingold JS, Baker A, Banks S, Baums IB, Cole J, Colgan MW, Fong P, Glynn PJ, Keith I, Manzello D, Riegl B, Ruttenberg BI, Smith TB, Vera-Zambrano M (2018) State of corals and coral reefs of the Galápagos Islands (Ecuador): past, present and future. Mar Pollut Bull 133:717–733. https://doi.org/10.1016/j.marpolbul.2018.06.002

González JA, Montes C, Rodríguez J, Tapia W (2008) Rethinking the Galapagos Islands as a complex social-ecological system: implications for conservation and management. Ecol Soc 13(2)

Greenslade P, Ireson J, Skarżyński D (2014) Biology and key to the Australian species of *Hypogastrura* and *Ceratophysella* (Collembola: Hypogastruridae). Austral Entomol 53(1):53–74

Grenier C (2007) Conservación contra natura. Las Islas Galápagos. Editorial Abya Yala

Grez AA, Zaviezo T, Orellana I, Pino L, Roy HE, Torres F, Rebolledo C (2022) Citizen science and Phytosanitary surveillance systems are complementary tools to follow the invasion of *Harmonia axyridis*. Neotrop Entomol. https://doi.org/10.1007/s13744-022-01014-0

Gurevitch J, Scheiner SM, Fox GA (2006) The ecology of plants, 2nd edn. Sinauer Associates

Hierro JL, Maron JL, Callaway RM (2005) A biogeographical approach to plant invasions: the importance of studying exotics in their introduced and native range. J Ecol 93(1):5–15

Hoddle MS, Ramírez CC, Hoddle CD, Loayza J, Lincango MP, Van Driesche RG, Causton CE (2013) Post release evaluation of *Rodolia cardinalis* (Coleoptera: Coccinellidae) for control of *Icerya purchasi* (Hemiptera: Monophlebidae) in the Galapagos Islands. Biol Control 67(2):262–274

Hoogmoed MS (1989) Introduced geckos in Puerto Ayora, Santa Cruz, with remarks on other areas

Hoskin CJ (2011) The invasion and potential impact of the Asian House Gecko (*Hemidactylus frenatus*) in Australia. Austral Ecol 36(3):240–251. https://doi.org/10.1111/j.1442-9993.2010.02143.x

Huber BA, Acurio AE (2022) Are introduced spiders displacing native species on Galápagos? Observations on pholcid spiders (Araneae: Pholcidae). Arachnology 19(sp1) https://doi.org/10.13156/arac.2022.19.sp1.191

IUCN (2017) IUCN Red List of Threatened Species: *Cedrela odorata*. IUCN Red List of Threatened Species. https://www.iucnredlist.org/en. Accessed 12 Dec 2022

James T, Vege S, Aldrich P, Hamrick JL (1998) Mating Systems of Three Tropical dry Forest Tree Species. Biotropica 30(4):587–594

Keith I, Dawson TP, Collins KJ, Campbell ML (2016) Marine invasive species: establishing pathways, their presence and potential threats in the Galapagos marine reserve. Pac Conserv Biol 22(4):377–377. https://doi.org/10.1071/pc15020

Keith I, Bensted-Smith W, Banks S, Suarez J, Riegl B (2022) Caulerpa chemnitzia in Darwin threatening Galapagos coral reefs. PLoS One 17(8):e0272581–e0272581. https://doi.org/10.1371/journal.pone.0272581

Kelley D, Page K, Quiroga D, Salazar R, Kelley D, Page K et al (2019) Geoconservation, geotourism and sustainable development in the Galapagos. In: In the footsteps of Darwin: Geoheritage, geotourism and conservation in the Galapagos Islands, pp 95–133

Kenis M, Auger-Rozenberg M-A, Roques A, Timms L, Péré C, Cock MJW, Settele J, Augustin S, Lopez-Vaamonde C (2009) Ecological effects of invasive alien insects. Biol Invasions 11(1):21–45. https://doi.org/10.1007/s10530-008-9318-y

Klein J, Verlaque M (2008) The *Caulerpa racemosa* invasion: a critical review. Mar Pollut Bull 56(2):205–225. https://doi.org/10.1016/j.marpolbul.2007.09.043

Kricher J, Loughlin K (2022) Galápagos: a natural history, 2nd edn. Princeton University Press

Lack D (1983) Darwin's finches. Cambridge University Press

Laso FJ, Benítez FL, Rivas-Torres G, Sampedro C, Arce-Nazario J (2020) Land cover classification of complex agroecosystems in the non-protected highlands of the Galapagos Islands. Remote Sens 12(1):65. https://doi.org/10.3390/rs12010065

Latorre O (1997) Galapagos: Los primeros habitantes de algunas islas

Lawesson J, Ortiz L (1987) Plantas introducidas en las Islas Galápagos. In: Investigación Botánica y Manejo en Galápagos: Memorias (Versión en Español) Taller sobre Investigación Botánica y Manejo en Galápagos. USAID, pp 224–235

Le Roux J, Wieczorek AM (2009) Molecular systematics and population genetics of biological invasions: towards a better understanding of invasive species management. Ann Appl Biol 154(1):1–17. https://doi.org/10.1111/j.1744-7348.2008.00280.x

Lee CE, Gelembiuk GW (2008) Evolutionary origins of invasive populations. Evol Appl 1(3):427–448

Letschert J, Wolff M, Kluger LC, Freudinger C, Ronquillo J, Keith I (2021) Uncovered pathways: modelling dispersal dynamics of ship-mediated marine introduced species. J Appl Ecol 58(3):620–631. https://doi.org/10.1111/1365-2664.13817

Levi HW (1968) The spider genera *Gea* and *Argiope* in America (Araneae: Araneidae). Bull Museum Comparat Zool 136(9):319–352

Levi HW (1993) American Neoscona and corrections to previous revisions of Neotropical orbweavers (Araneae: Araneidae). Psyche J Entomol 99:221–239

Levi HW (2004) Comments and new records for the American genera *Gea* and *Argiope* with the description of a new species (Araneae: Araneidae). Bull Museum Comparat Zool 158(2):47–65

Lundh JP (2001) Galápagos: a brief history. http://www.lundh.no/jacob/galapagos/pg05.htm#BIBLIOGRAPHY. Accessed 10 Nov 2022

Lundh JP (2006) Farm area and cultivated plants on Santa Cruz, 1932-1965, with remarks on other parts of Galapagos. Galapagos Res 64:12–25

Mauchamp A (1997) Threats from alien plant species in the Galápagos Islands. Conserv Biol:260–263

Mauchamp A, Atkinson R (2009) Rapid, recent and irreversible habitat loss: *Scalesia* forest on the Galapagos Islands. Galapagos Report 2009–2010. GNPD, GCREG, CDF and GC. Puerto Ayora, Galapagos, Ecuador 2010: 108–112

McMullen CK (1999) Flowering plants of the Galápagos. Cornell University Press

McNew SM, Clayton DH (2018) Alien invasion: biology of *Philornis* flies highlighting *Philornis downsi*, an introduced parasite of Galápagos birds. Annu Rev Entomol 63:369–387

Meyer JY, Fourdrigniez M, Veitch CR, Clout MN, Martin AR, Russell JC, West CJ (2019) Islander perceptions of invasive alien species: the role of socio-economy and culture in small isolated islands of French Polynesia (South Pacific). Island Invasives Scaling Up Meet Challenge 62(510)

Miura O (2007) Molecular genetic approaches to elucidate the ecological and evolutionary issues associated with biological invasions. Ecol Res 22(6):876–883. https://doi.org/10.1007/s11284-007-0389-5

Nuez F, Prohens J, Blanca JM (2004) Relationships, origin, and diversity of Galapagos tomatoes: implications for the conservation of natural populations. Am J Bot 91(1):86–99

Oleas-Paz MP, Cisneros-Heredia DF (2019) Spatial distribution and microhabitat use of species of geckos in San Cristobal Island, Galapagos. Universidad San Francisco de Quito, Quito

Olmedo J, Cayot LJ (1994) Introduced geckos in the towns of Santa Cruz, San Cristóbal and Isabela. Noticias de Galapagos 53(7):7–12

Oxford P, Watkins G, Philip P (2009) Galapagos: both sides of the coin

Peck SB (2006) The beetles of the Galápagos Islands, Ecuador: evolution, ecology, and diversity (Insecta: Coleoptera). NRC Research Press

Pennington TD, Muellner AN (2010) Monograph of *Cedrela* (Meliaceae). dh books

Phillips RB, Wiedenfeld DA, Snell HL (2012) Current status of alien vertebrates in the Galápagos Islands: invasion history, distribution, and potential impacts. Biol Invasions 14(2):461–480. https://doi.org/10.1007/s10530-011-0090-z

Quinton JM, Fay MF, Ingrouille M, Faull J (2011) Characterisation of *Rubus niveus*: a prerequisite to its biological control in oceanic islands. Biocontrol Sci Tech 21(6):733–752. https://doi.org/10.1080/09583157.2011.570429

Quiroga D (2009) Crafting nature: the Galapagos and the making and unmaking of a "natural laboratory". J Polit Ecol 16(1):123–140

Quiroga D (2013) Changing views of the Galapagos. In: Walsh SJ, Mena CF (eds) Science and conservation in the Galapagos Islands. Springer, New York, pp 23–48

Ramos-Rojas SA, Cisneros-Heredia DF (2019) Morfología comparada y ecología de las comunidades de gecos en la isla San Cristóbal, Galápagos y el continente ecuatoriano. Universidad San Francisco de Quito, Quito

Reatini B, De Lourdes Torres M, Vision TJ (2022) Local exclusion and regional decline of an endemic Galápagos tree species (*Psidium galapageium*) by an invasive relative (*P. guajava*). https://doi.org/10.1101/2022.10.11.511772

Renteria JL, Gardener MR, Panetta FD, Crawley MJ (2012) Management of the invasive hill raspberry (*Rubus niveus*) on Santiago Island, Galapagos: eradication or indefinite control? Invasive Plant Sci Manag 5(1):37–46

Rhoades OK, Brandt M, Witman JD (2023) La Niña-related coral death triggers biodiversity loss of associated communities in the Galápagos. Mar Ecol 44(5):e12767. https://doi.org/10.1111/maec.12767

Riegl B, Glynn PW, Banks S, Keith I, Rivera F, Vera-Zambrano M et al (2019) Heat attenuation and nutrient delivery by localized upwelling avoided coral bleaching mortality in northern Galapagos during 2015/2016 ENSO. Coral Reefs 38:773–785. https://doi.org/10.1007/s00338-019-01787-8

Rivas-Torres G, Rivas MG (2018) Allelopathic Impacts of the Invasive Tree *Cedrela odorata* L.(Meliaceae, Sapindales= Magnoliidae) in the Galapagos Flora. Understanding invasive species in the Galapagos Islands: From the molecular to the landscape, 77–93

Rivas-Torres G, Luke Flory S, Loiselle B (2017) Plant community composition and structural characteristics of an invaded forest in the Galápagos. Biodivers Conserv 27(2):329–344. https://doi.org/10.1007/s10531-017-1437-2

Rivas-Torres GF, Benítez FL, Rueda D, Sevilla C, Mena CF (2018) A methodology for mapping native and invasive vegetation coverage in archipelagos: an example from the Galápagos Islands. Prog Phys Geography Earth Environ 42(1):83–111

Rowe JW, Martin CE, Clark DL, Valle C, Vintimilla Palacios CP (2019) Habitat use and spatial ecology of three *Microlophus* lizard species from Santa Cruz and San Cristóbal Islands, Galápagos, and the coastal dry forest of Machalilla, Ecuador. Herpetol Rev 50(1):43–51

Sakai AK, Allendorf FW, Holt JS, Lodge DM, Molofsky J, With KA et al (2001) The population biology of invasive species. Annu Rev Ecol Syst:305–332. https://doi.org/10.1146/annurev.ecolsys.32.081501.114037

Sampedro C, Pizzitutti F, Quiroga D, Walsh SJ, Mena CF (2020) Food supply system dynamics in the Galapagos Islands: agriculture, livestock and imports. Renew Agric Food Syst 35(3):234–248

Schaffelke B, Smith JE, Hewitt CL (2006) Introduced macroalgae – a growing concern. J Appl Phycol 18(3–5):529–541. https://doi.org/10.1007/s10811-006-9074-2

Silberglied RE (1978) Inter-Island transport of insects aboard ships in the Galápagos Islands. Biol Conserv 13(4):273–278

Sturaro MJ, Avila-Pires TC (2013) Redescription of the gecko Gonatodes caudiscutatus (Günther, 1859) (Squamata: Sphaerodactylidae). S Am J Herpetol 8(2):132–145

Swanepoel L, Loyola A, Flowers J, Lewbart GA, Garrett KB, Yabsley MJ (2022) Infection with a novel pentastome (*Raillietiella sp.*) in a juvenile endangered Galápagos tortoise (*Chelonoidis Niger*). Vet Parasitol Regional Stud Rep 35:100782

Toral-Granda MV, Causton CE, Jäger H, Trueman M, Izurieta JC, Araujo E, Cruz M, Zander KK, Izurieta A, Garnett ST (2017) Alien species pathways to the Galapagos Islands, Ecuador. PLoS One 12(9):e0184379. https://doi.org/10.1371/journal.pone.0184379

Torres M, Mena CF (2018) Understanding invasive species in the Galapagos Islands. Springer, Cham

Torres ML, Urquía D, Moyle L, Gibson M, Vision T, Reatini B (2023) Unravelling the interactions between endemic and invasive plant species in the Galapagos Islands. In: Island ecosystems: challenges to sustainability. Springer, pp 169–186

Torres-Carvajal O, Tapia W (2011) First record of the common house gecko Hemidactylus frenatus Schlegel, 1836 and distribution extension of *Phyllodactylus reissii* Peters, 1862 in the Galápagos. 2009–2011

Torres-Carvajal O, Barnes CW, Pozo-Andrade MJ, Tapia W, Nicholls G (2014) Older than the islands: origin and diversification of Galápagos leaf-toed geckos (Phyllodactylidae: *Phyllodactylus*) by multiple colonizations. J Biogeogr 41(10):1883–1894

Torres-Carvajal O, Rodríguez-Guerra A, Chaves JA (2016) Present diversity of Galápagos leaf-toed geckos (Phyllodactylidae: *Phyllodactylus*) stems from three independent colonization events. Mol Phylogenet Evol 103:1–5

Tye A (2006) Can we Infer Island introduction and naturalization rates from inventory data? Evidence from introduced plants in Galapagos. Biol Invasions 8(2):201–215. https://doi.org/10.1007/s10530-004-3574-2

Urquía D, Gutierrez B, Pozo G, Pozo MJ, Espín A, de Torres ML (2019) *Psidium guajava* in the Galapagos Islands: population genetics and history of an invasive species. PLoS One 14(3):e0203737. https://doi.org/10.1371/journal.pone.0203737

Urquía D, Pozo G, Gutierrez B, Rowntree JK, de Torres ML (2020) Understanding the genetic diversity of the guayabillo (*Psidium galapageium*), an endemic plant of the Galapagos Islands. Global Ecol Conserv 24:e01350. https://doi.org/10.1016/j.gecco.2020.e01350

Urquía D, Gutierrez B, Pozo G, Pozo MJ, Torres MDL (2021) Origin and dispersion pathways of guava in the Galapagos Islands inferred through genetics and historical records. Ecol Evol 11(21):15111–15131. https://doi.org/10.1002/ece3.8193

Valdebenito H (2018) A study contrasting two congener plant species: *Psidium guajava* (introduced guava) and *P. Galapageium* (Galapagos guava) in the Galapagos Islands. In: de Torres ML, Mena CF (eds) Understanding invasive species in the Galapagos Islands: from the molecular to the landscape. Springer International Publishing, pp 47–68. https://doi.org/10.1007/978-3-319-67177-2_4

Venegas P, Calderon F, Riofrío D, Benítez D, Ramón G, Cisneros-Heredia D, Coimbra M, Rojo-Álvarez JL, Pérez N (2021) Automatic ladybird beetle detection using deep-learning models. PLoS One 16(6):e0253027. https://doi.org/10.1371/journal.pone.0253027

Vincek V, O'Huigin C, Satta Y, Takahata N, Boag PT, Grant PR et al (1997) How large was the founding population of Darwin's finches? Proc R Soc Lond Ser B Biol Sci 264(1378):111–118

Ward SM, Gaskin JF, Wilson LM (2008) Ecological genetics of plant invasion: what do we know? Invasive Plant Sci Manag 1(1):98–109. https://doi.org/10.1614/IPSM-07-022.1

Wauters N, Dekoninck W, Hendrickx F, Herrera HW, Fournier D (2016) Habitat association and coexistence of endemic and introduced ant species in the Galápagos Islands. Ecol Entomol 41(1):40–50

Williams SL, Grosholz ED (2002) Preliminary reports from the *Caulerpa taxifolia* invasion in southern California. Mar Ecol Prog Ser 233:307–310. https://doi.org/10.3354/meps233307

Witman JD, Smith F (2003) Rapid community change at a tropical upwelling site in the Galápagos marine reserve. Biodivers Conserv 12(1):25–45

Wolf JB, Tautz D, Trillmich F (2007) Galápagos and Californian sea lions are separate species: genetic analysis of the genus *Zalophus* and its implications for conservation management. Front Zool 4:1–13

Wright JW (1983) The evolution and biogeography of the lizards of the Galápagos archipelago: evolutionary genetics of *Phyllodactylus* and *Tropidurus* populations. In: Patterns of evolution in Galápagos organisms, pp 123–155

Chapter 8
Supporting Galápagos Native Species via Ecological Landscape Design in Urban Greenspaces

Gail A. Langellotto, Diego F. Cisneros-Heredia, Mateo Dávila-Jativa, Sohyun Park, Nina M. Emond Miller, Daniel Velasco-Cedeño, Wenshu Sun, Emilia Peñaherrera-Romero, Ariel Guerrero-Campoverde, Roberto J. León-E, Fei Mo, Daniel Die Morejón, and Hui Li

Abstract Like other regions of the world, urban gardens in the Galápagos are dominated by non-native plants, several of which have escaped cultivation and become naturalized or invasive. The negative environmental and economic consequences of escape and invasion are particularly high in island systems, where insular endemics may be at increased risk of local or global extinction. Landscape designs that prominently feature native plants and favor local biodiversity are increasingly popular, yet largely unavailable for the Galápagos. The potential benefits of native plant landscape designs are considerable. Specifically, native plant landscape designs reduce or eliminate the potential for plant invasions and provide host plants and habitat for native wildlife. In this chapter, we discuss current socio-economic concerns and regulatory constraints that guide or constrain design standards for gardens on the Galápagos, and the potential for incorporating native plants into garden designs that favor native wildlife. We share a list of recommended plant materials for landscape designs on the island of San Cristóbal, including those with that benefit pollinators. We include guidance that can be used in ornamental landscape designs for island systems for supporting native plant gardens proliferation on San Cristóbal, and throughout the Galápagos, in general.

G. A. Langellotto (✉) · N. M. Emond Miller
Department of Horticulture, Oregon State University, Corvallis, OR, USA
e-mail: gail.langellotto@oregonstate.edu

D. F. Cisneros-Heredia · M. Dávila-Jativa · D. Velasco-Cedeño · E. Peñaherrera-Romero · A. Guerrero-Campoverde · R. J. León-E · D. D. Morejón
Universidad San Francisco de Quito Tropical Biodiversity Institute, Quito, Ecuador

S. Park · H. Li
Department of Plant Science and Landscape Architecture, University of Connecticut, Mansfield, CT, USA

W. Sun · F. Mo
Shanghai Jiaotong, University School of Design, Shanghai, China

J. Dunn Insua et al. (eds.), *Human Galapagos*, Social and Ecological Interactions in the Galapagos Islands, https://doi.org/10.1007/978-3-032-08800-0_8

Introduction

Island systems host unique floral assemblages characterized by a high level of endemic species. On the Galápagos Islands, more than half the indigenous plant taxa are endemic, meaning they cannot be seen anywhere else on the globe (Stöcklin 2009). For example, the genus *Scalesia* (Family: Asteraceae) includes 15 endemic trees and shrub species that radiated from a common ancestor to exploit vacant niches in the Galápagos. These *Scalesia* species have adapted in fascinating ways, such as evolving drought-resistant leaves markedly different in shape and size (McMullen 1999). One species within the genus (*S. pedunculata)* grows in dense stands of tall, thin trees, forming a forest that shares taxonomic grouping with sunflowers, zinnias, and daisies. The rich and unique natural heritage of the Galápagos is recognized for its outstanding universal value, which drives tourism to the islands. Tourism started developing on the islands in the 1960s, and has greatly accelerated in the past few decades, from 40,000 in 1990 to 225,000 in 2016 (de Haan et al. 2019). Reports for 2017 are over 3,000,000 visitors (Mena et al. 2020). This growing influx of visitors has fundamentally changed the local economy, with a shift away from farming and fishing in favor of tourism (Walsh et al. 2019).

The socio-economic shift has facilitated large-scale landscape change. Native plants are defined as those that were present prior to human settlement (Iannone et al. 2020), which occurred in the 1890s in the Galápagos with the establishment of the El Progresso settlement. There are now over 800 non-native plant species on the islands, compared to about 550 native plant taxa (Gardener et al. 2013; Trueman et al. 2010; Toral-Granda et al. 2017). Most of these non-native plants (about 700) were deliberately introduced for food or ornamental purposes, representing the most significant pathway for the intentional introduction of non-native species in the Galápagos (Tallamy et al. 2010; Toral-Granda et al. 2017). However, unintentional introductions also occur, with an estimated 130 species arriving as seeds or plant-associated material (Guézou et al. 2010; Toral-Granda et al. 2017). As many as one-third of these non-native plants have naturalized outside of areas of cultivation, becoming invasive species that threaten the Galápagos' unique biodiversity (Trueman et al. 2010). Although agricultural activities facilitated non-native plant introductions, farm management practices like weeding also helped to keep these species in check (Giefer 2023). However, as some farmlands in the rural highlands are being abandoned to pursue a tourism-based economy, many non-native plants are escaping cultivation and establishing in surrounding wildlands.

In the urban lowlands, non-native plant introductions are largely associated with residential gardens, public parks, and commercial landscaping. As tourist activities increased on the island, so did residential and commercial construction, leading to a surge in associated landscaping projects. For example, hotel construction boomed nearly five-fold between 2006 and 2017, from 65 to 317 hotels (Robertson 2022). Ornamental plants, popular for landscaping, represent more than half of the non-native flora across the Galápagos, with an even higher representation (58%) of non-native flora in urban areas (Guézou et al. 2010). When non-native flora escapes cultivation and causes economic or ecological harm, they are considered invasive (Iannone et al. 2020). An estimated 14% of ornamental plants have escaped cultivation and naturalized in the surrounding environment, with some becoming invasive across the archipelago (e.g. *Lantana camara*) (Guézou et al. 2010). Other non-native, ornamental plants that are present in the Galápagos were note reported to have escaped cultivation in Guézou et al. 2010 (e.g. *Spathodea campanulata*, *Arundo donax*, *Leucaena leucocephala*) but are listed among the 100 worst invasive species on a global basis (Lowe et al. 2000). More recent observations of *Leucaena leucocephala*, however, suggest that this shrub has now escaped cultivation and into the unmanaged landscapes in the rural highlands (iNaturalist 2025). Thus, from both the rural highlands and the urban lowlands, the unique and historically significant ecosystems of the Galápagos are being reshaped by non-native, naturalized, and invasive plants.

Recognizing the potential threats posed by invasive non-native species, the Ecuadorian government took a multi-pronged approach to protect the Galápagos. In 1999, biosecurity protocols were established, making it illegal to bring new plant species to the archipelago (Guézou et al. 2010). These protocols aimed to prevent not only the intentional arrival of new plants often carried as cuttings or seedlings brought from mainland Ecuador gardens and agricultural areas, but also the unintended introduction of associated organisms like insects, pathogens, and even other plants (Toral-Granda et al. 2017). Building on these efforts, a strategic plan for invasive species management was developed in 2017. In 2012, the Galápagos Biosecurity Agency (ABG) was formed to address the challenge of preventing and reducing the introduction of non-native species, safeguarding human health, biodiversity, and the ecological integrity of the islands.

Despite these measures, challenges remain. While studies by Toral-Granda et al. (2017) suggest a decrease in plant confiscations from Galápagos residents (during biosecurity inspections) after a decade of the establishment of biosecurity policies, introductions still occur. For example, commercial, edible *Passiflora* fruits are repeatedly introduced by residents and tourists (i.e. 479 interceptions in 2016, Toral-Granda et al. 2017), even though *Passiflora ligularis* has long been naturalized in the Galápagos (Guézou et al. 2010). Enforcing quarantines and other biosecurity protocols across the vast archipelago continues to be a complex task. Likewise, eradication programs have had limited success. Few plant eradication projects have been successful in the Galápagos, usually those targeting small areas (no more than 1 hectare in net area) and managed by a single owner without a persistent seed bank (Gardener et al. 2013).

The Galápagos Islands thus face a conundrum: the iconic natural beauty of the islands fuel ecotourism, but tourism is directly or indirectly responsible for conservation threats to native and endemic taxa. Strict enforcement challenges for introduction of non-native species and the high cost and complexity of eradication programs highlight the need for additional strategies. We propose a supporting approach: centering the use of native plants in urban greenspaces, including gardens and parks, and generating landscape designs that are biologically and culturally relevant. Transitioning urban greenspaces towards native plant landscapes can raise tourists' knowledge, awareness, and appreciation of the islands' native flora. Native plant designs can also prevent future plant invasions by offering cost-efficient ornamental alternatives and can better support local wildlife compared to non-native plantings. Native plant designs can feature and celebrate the unique natural heritage of the islands and promote conservation in an urban context. In this chapter, we will share a list of native flora that have good potential for use in urban landscape designs and guidelines for best practices in using native plants in urban landscapes, particularly for island systems.

The Case for Native Plants in Urban Greenspaces

An appreciation for nature defines one of the three primary market segments of Galápagos tourists (Carvache-Franco et al. 2021). The nature-appreciating tourist market segment also had the highest education level and spent the most money per day on the islands, relative to other categories of Galápagos tourists (Carvache-Franco et al. 2021). With the advancement of sustainable tourism concepts, there is an opportunity to highlight the beauty and ecosystem value of native plants in the built environment of eco-tourism destinations. Native plants not only exhibit strong ecological adaptability and reduce maintenance costs but also provide opportunities for ecological education. An analysis of tourists' perceptions of native plants at popular sites across the Galápagos (Sun and Mo, in prep) found that tourists value: (1) the ecological authenticity conveyed by native vegetation in its natural state; (2) the ecological interaction between native plants and animals, and (3) the interpretive role of plants in guided tours and environmental education. Tourist infrastructure (e.g. hotels, restaurants, and other commercial services) with native plant landscapes thus have an opportunity to attract a high-value segment of the Galápagos tourist market via the use of messaging that aligns with and reinforces visitors' intentions for an ecologically authentic experience.

Native plant landscape designs may also benefit local wildlife, including insects. About 50% of all insects are herbivorous, yet only <10% of insect species are considered true generalists (Bernays and Graham 1988), capable of feeding on plants across three or more families. The chemicals within plant leaves, stems, and other vegetative structures appear to prevent insect herbivores from easily switching to feeding on plants with novel chemistries (Winkler and Mitter 2008). Without a long-term, evolutionary relationship with a plant, an herbivorous insect may not be able to digest, detoxify, or recognize non-native plants as a suitable host (Jaenike 1990). Even among generalist herbivorous insects with a broad host plant range, many cannot survive and grow, when fed non-native plant taxa (Tallamy et al. 2010). Urban greenspaces that are predominantly landscaped with native plants thus support a higher abundance of herbivorous insects (Burghardt et al. 2010; Salisbury et al. 2017) and support more species of herbivorous insect taxa (Burghardt et al. 2010; Clem and Held 2015; Mata et al. 2021). Importantly, many researchers have found that conservation of herbivorous insect richness, biomass, and abundance scales up to the conservation of birds and other charismatic megafauna (Burghardt et al. 2009; Narango et al. 2017). Thus, choosing native plant-focused designs can initiate a bottom-up trophic cascade, where primary producers (native plants) mediate an increase of energy flow to higher trophic levels.

The benefits of native plants to pollinators in urban areas are somewhat less clear cut. Compared to herbivorous insects, pollinators have a different relationship with plants. Whereas vegetative plant parts have defenses to deter herbivore feeding, plant floral traits likely evolved to attract pollinators. In addition, unlike herbivorous insects, most insect pollinators tend to be generalists (Fontaine et al. 2009). In some cases, garden spaces landscaped with native plants did not significantly increase the abundance and/or diversity of pollinating insects, compared gardens planted with non-native plants (Matteson and Langellotto 2011; Majewska et al. 2018). However other studies have found that pollinator abundance and/or richness increase with native plant abundance in gardens (Pardee and Philpott 2014; Fukase and Simons 2016).

In the United States and Europe, interest in native plant gardening has drastically increased over the past few decades, in part because of efforts to provide habitat for wildlife (McMahan 2006). Native plant gardening, or gardening in general, isn't a popular pastime in the Galápagos. However, recent programs, including a native plant garden installation at a luxury hotel, are working to increase their popularity (Atkinson 2008; Atkinson et al. 2017; Diaz 2019). The ability to see endemic organisms (mostly animals) and general interests in biodiversity and conservation are important drivers of tourism to the Galápagos islands (Mazur et al. 2018). Although native plant tourism is not currently a strong draw for most tourists, local officials and businesses could market efforts to conserve the islands' unique natural history via native plant landscapes, to positive economic outcomes. Studies in India have demonstrated that visitors are willing to pay a modest premium to support local native plant projects (Sardana 2019). Thus, there is some precedence for mobilizing tourist revenue to support native plantings.

Although relatively rare, there are examples of small-scale, successful plantings that utilize native plants along the urban settlements in the archipelago. In the "malecón" (boardwalk) of Puerto Ayora and Puerto Baquerizo Moreno, a diverse set of native plants are present as ornamental decoration, including the Galápagos Pricklypear (*Opuntia galapageia*), Candelabra Cactus (*Jasminocereus thouarsii*), Scalesias (*Scalesia spp.*), Yellow Geiger or "muyuyo" (*Cordia lutea*), Darwin's Aster (*Darwiniothamnus spp.*) and many others. Another important example is the Universidad San Francisco de Quito- Galápagos Academic Institute for the Arts and Sciences (USFQ-GAIAS) campus extension in San Cristóbal. The gardens around the facilities host native plants, like the Curve-spined Lecocarpus (*Lecocarpus darwinii*), and some cultivation systems. These compositions act as an ideal framework that should be applied to the rest of urban environments in Galápagos human settlements.

Native Plants for Galápagos Gardens and Greenspaces

The Galápagos Islands are listed as the top world heritage in the natural property category by UNESCO in 1973, thereafter its Outstanding Universal Values (OUV) in biological diversity conservation has been recognized globally (O'Donnell 2004). Planting design has a crucial relationship to the OUV of the world heritage sites, and this has been emphasized as a key criterion for assessing the OUV. According to the 2023 version of the Operational Guidelines for the Implementation of the World Heritage Convention, being an outstanding example representing the significant ecological and biological processes and the development of communities of plants is listed as one of the criteria for the assessment of OUV (UNESCO 2023). Therefore, native plants usage and design in the Galápagos will benefit the recognition of its OUV from the World Heritage Conservation aspect, as it improves public comprehension of native plants and local identity of the sites and will facilitate human-nature interactions locally.

To support future work focused on the use of native plants in urban gardens and greenspaces, we developed a list of Galápagos native and endemic plant species with ornamental qualities. Our list expands upon the excellent work of Atkinson et al. (2017), who provide a list of 89 native and endemic species for Galápagos gardens, written in Spanish, Kichwa (indigenous Ecuadorean language), and English. Our list includes 63 species (Table 8.1) and is coupled with a sample landscape design (Fig. 8.1). We focused on plants that were native or endemic to the island of San Cristóbal, although most plants could be found across populated islands in the Galápagos archipelago. We drew heavily from the work of McMullen (1999), with special attention to plants with noted ornamental qualities, including colorful flowers or fruit, or other visually appealing traits. Plants mentioned in McMullen (1999) but without detailed plant characteristics (e.g. form, height, branch, leaf, flower, fruit, seed) were described from published monographs by Burkart (1976), Kunze (1847), and Smith and Tejero-Díez (2014), and identification keys by Atkinson et al. (2017), Beentje et al. (2005), Mickel and Beitel (1988), Skolmen (n.d.), Smith (1987), and Verdcourt (1992).

Our list does not include plants listed as endangered (e.g. *Scalesia divisa, Scalesia gordillo, Calandrinia galápagosa, Encilia hispida*), vulnerable (e.g. *Alternanthera flavicoma*, *Froelichia nudicaulis*, *Scalesia incisa*, *Scalesia pedunculata*, *Tournefortia rufo-sericea*, *Miconia robinsoniana*), or near threatened (*Justicia galapagana*, *Lithophila radicata*, *Lecocarpus darwinii*, *Gossypium klotzschianum*) on the IUCN Red List (2023). Propagation and distribution of these plants should be managed for restoration and conservation, rather than landscaping projects. Note, however, that some of the species (e.g. *Scalesia* and *Leuocarpus*) are planted along the malecón (boardwalk) and in the gardens of government buildings (Velasco 2025, personal observation). We also removed plants with thorns or spines (e.g. *Zanthoxylum fagara*, *Castela galapageia*, *Scutia spicata*, *Acacia insulae-iacobia*, with the notable exceptions of the candelabra cactus (*Jasminocereus thouarsii*) and the prickly pear cactus (*Opuntia megasperma*) which are already in wide use in San

Table 8.1 Plants native or endemic to the Galápagos Islands with ornamental features for urban Galápagos landscape designs

Layer (after Rainer and West 2015)	Family	Scientific name	Common name(s)	Range	Habitat	Notes	Height	Leaves	Flowers	Fruit
Seasonal theme plant	Asteraceae	*Pectis subsquarrosa* **	Pectis, Estrella dorada	E	Arid lowlands	Distinct aroma when leaves are rubbed between the fingers. Suitable for rock gardens.	50 cm tall	Simple, oblong leaves, 0.4–2.5 cm long, covered with oil glands.	Radiate heads that are yellow from above and reddish brown from beneath.	Achene, black, 2.5–3.5 mm long.
Seasonal theme plant	Caesalpiniaceae	*Senna uniflora*	Rufous-haired senna	N	Arid lowlands	Similar to non-native *Senna obtusifolia*,but possesses reddish brown hairs covering most of the plant.	1 m tall	Alternate, even-pinnately compound; leaflets 4–6, oval to obovate, tips somewhat rounded; margins fringed with tiny hairs	Solitary or in axillary racemes; yellow, 5 petals 9–12 mm long	Legume, curved and 4-angled, 10–20 cm long
Filler	Molluginaceae	*Mollugo flavescens*	Mollugo	E	Arid lowlands	A slight, delicate flower that is well suited to rock gardens.	40 cm tall	Whorls, simple; blade linear, oblanceolate, spatulate, or obovate, 5–20 mm long, smooth; margins entire	Axillary, typically 3–8 per cluster; sepals whitish with green veins 1.5–2 mm long	Capsule
Seasonal theme plant	Polygalaceae	*Polygala galapageia*	Galápagos milkwort	E	Arid lowlands	Well suited for sandy sites, such as plantings adjacent to an urban boardwalk.	1 m tall	Alternate, simple; blade oblanceolate to spatulate, 6–25 mm long, leathery, with bluish, waxy covering	Terminal racemes to 12 cm long; 5 nerved; purplish to white	Capsule, oblong; 3.5–4 mm long
Seasonal theme plant	Solanaceae	*Physalis angulata*	Ground cherry, Cutleaf ground cherry	N	Arid lowlands	Lantern-like husks cover fruit. When ripe, small fruits are edible, and valued for unique flavor and medicinal properties. Can be grown in pots.	90 cm tall	Dark green and roughly oval, often with tooth shapes around the edge.	Axillary, solitary; corolla yellow, campanulate with 5 lobes, 7–10 mm long; anthers bluish	Berry, round, 10–18 in diameter, enclosed in a 10-angled calyx

(continued)

Table 8.1 (continued)

Layer (after Rainer and West 2015)	Family	Scientific name	Common name(s)	Range	Habitat	Notes	Height	Leaves	Flowers	Fruit
Seasonal theme plant	Solanaceae	*Physalis cordata*	Galápagos ground cherry, heartleaf ground cherry	E	Arid lowlands	Lantern-like and attractive husks cover fruit. When ripe, small fruits are edible, and valued for unique flavor and medicinal properties. Can be grown in pots.	90 cm tall	Large, heart-shaped leaves, 4–9 cm long; surfaces somewhat hairy; margins irregularly toothed, occasionally entire	Axillary, solitary; corolla yellow, campanulate with 5 lobes, 7–10 mm long; anthers bluish	Berry, round, 10–18 in diameter, enclosed in a 5-angled calyx, 30–45 mm long and 37–40 mm across; smooth
Filler	Oxalidaceae	*Oxalis dombeyi*	Dombey's wood sorrel	N	Arid lowlands and moist uplands	Good accent plant for garden beds and containers. Leaves and flowers are edible.	50 cm tall	Shamrock-like leaves, pinnately compound with 3 leaflets, margins with fine hairs	Axillary, umbel-like cymes; yellow, sometimes with orange along veins; petals 5.5–7 mm long	Capsule, oblong, 7–10 mm long
Filler	Asteraceae	*Blainvillea dichotoma*	Blainvillea	N	Arid lowlands, moist uplands	In the wild, found in open or disturbed sites. May be planted in high-traffic areas, as a filler plant.	1 m tall	Blade lanceolate to broadly ovate, 1.5–8 cm long, margins variously serrate	Small, sparse white flowers. Radiate heads, solitary, terminal, axillary.	Achene, black, 2–3.5 mm long with 3 awns on top to 1.5 mm long
Filler	Aizoaceae	*Trianthema portulacastrum*	Trianthema	N	Coastal	Can be grown in rock gardens and other challenging sites. Seeds can be dispersed by water. Spreads well, and can be invasive outside of N range.	Low-growing	Opposite, 1 of each pair larger in size, bases meeting at stems to form a tiny cup, simple; blade obovate or spatulate, 1–2 cm long, margins entire	Axillary, solitary; white or purple, 5 petals ~5 mm long, corolla absent	Capsule, ovoid with a caplike lid

Seasonal theme plant	Solanaceae	*Exedeconus miersii*	Galápagos shore petunia	E	Coastal	Occasionally seen inhabiting arid lowlands, but most often seen growing along rocky or sandy shorelines. Drought-tolerant.	Low-growing	Large (4–20 cm long), hairy light green leaves that form a dense ground cover. Margins irregularly toothed	White, trumpet-shaped flowers with green veins. Axillary, solitary, sometimes appearing clustered	Capsule, elongate-ovate
Filler	Asteraceae	*Jaegeria gracilis*	Galápagos jaegeria	E	Moist uplands	Slender and trailing annual with small, sessile flowers. In the wild, it can be found in moist soil at disturbed sites and in the fern-sedge or grassy zone.	30 cm tall	Opposite, simple; blade lanceolate to ovate, 1–4 cm long; margins usually entire	Solitary, yellow, radiate flowerheads, 4–6 mm across.	Achene, black, 1 mm long
Filler	Asteraceae	*Eclipta alba*	False daisy	N	Moist uplands	Slender annual with delicate, white flowers. May be planted and maintained as a living mulch around trees.	1 m tall	Opposite, simple; blade lanceolate, elliptic-lanceolate, linear-lanceolate, 2–10 cm long; margin minutely serrate	White, radiate heads, 6 mm across. Blooms are solitary or in groups of 2–3.	Achene, 2–2.5 mm long, somewhat warty

(continued)

Table 8.1 (continued)

Layer (after Rainer and West 2015)	Family	Scientific name	Common name(s)	Range	Habitat	Notes	Height	Leaves	Flowers	Fruit
Seasonal theme plant	Boraginaceae	*Heliotropium indicum #*	Indian heliotrope, cola de escorpion	N	Moist uplands	Prefers open areas with a mixture of sunshine and shade. Intermediate in height/form, compared to the other two of this genus. Could be potted and maintained in arid lowlands with irrigation.	1 m tall	Alternate, simple; blade ovate to broadly lanceolate, 3–15 cm long; hairy, magins entire to slightly waxy	Scorpioid spike; purplish or bluish with white and/or yellow throat, occasionally entirely white; funnelform 3–5 mm long	4 nutlets, 2.5–3.5 mm long, with pointed tip and 2–3 lines on the back
Filler	Caryophyllacceae	*Drymaria cordata*	Drymaria	N	Moist uplands	Young leaves are edible. In arid lowlands, may be grown in containers. Outside of N range, can spread aggressively.	Low-growing	Opposite, simple; blade reniform to roundish, 8–20 mm long, 10–20 mm wide, margins entire	Terminal or axillary cymes; white, 5 petals 2–2.5 mm long, each petal deeply 2 lobed	Capsule

Herbaceous perennials										
Seasonal theme plant	Asteraceae	*Pectis tenuifolia*	Pectis	E	Arid lowlands	Found growing in volcanic areas, in the wild. Would be an excellent specimen in rock gardens. Distinct aroma when leaves are rubbed between the fingers.	50 cm tall	Opposite, simple; lade linear-oblong, 0.4–2.5 cm long; lower surface with distinct oil glands; margins entire except near base where there may be a few pairs of stiff bristles	Radial flowers are yellow from above and red from beneath. Flower heads are 6–7 mm across, borne near stem tips.	Achene, black, 2.5–3.5 mm long, with 2 to many scales or bristles on top, 2–3 mm long, sometimes no scales or bristles
Seasonal theme plant	Boraginaceae	*Heliotropium angiospermum* ** #	Heliotrope, cola de escorpion	N	Arid lowlands	Beautiful scorpion's tail form of flowers. Not to be confused with *Heliotropium anderssonii* (vulnerable endemic only known from Santa Cruz) or *Heliotropium rufipilum* var. *anademum* (an introduced species found on Floreana).	50 cm tall	Alternate, simple; blade elliptical, ovate, sometimes lanceolate, 1.5–10 cm long, 1–2 cm; with white hairs	Scorpioid spikes to 20 cm long; white, often with yellowish green throat; campanulate with 5 lobes, 1.5–2 mm long, throat hairy	Nutlets, 2–2.8 mm across

(continued)

Table 8.1 (continued)

Layer (after Rainer and West 2015)	Family	Scientific name	Common name(s)	Range	Habitat	Notes	Height	Leaves	Flowers	Fruit
Seasonal theme plant	Caesalpiniaceae	*Senna occidentalis* ** #	Coffee senna	N	Arid lowlands	In the wild, appears to favor open, disturbed areas, sprawling herb. May be a good candidate for high traffic areas. Not to be confused with *Senna hirsuta* var. *hirsuta*, an introduced species with a foul odor.	2 m tall	Alternate, even-pinnately compound; leaflets 8–12 (usually 10), elliptic, ovate or lanceolate, sharply pointed, 6 cm long, margins with tiny hairs	Axillary or terminal racemes; yellow, petals 12–15 mm long	Legume, linear, 8–10 cm long
Filler	Malvaceae	*Bastardia viscosa*	Bastardia	N	Arid lowlands	Often found growing in areas with abundant sunlight around shrubs	1 m tall	Alternate, simple; blade ovate to somewhat cordate, to 10 cm long; margins entire or with small teeth	Axillary, solitary, or in threes; yellow to yellowish orange; petals 5–6 mm long, stamens united in a tubular column	Schizocarp, roundish, 6–7 mm across, beaked, usually with 6–7 sections
Seasonal theme plant	Solanaceae	*Lycopersicon cheesmanii*	Galápagos tomato, tomatillo	E	Arid lowlands	Sprawling plant found on rough lava areas and disturbed sites. Fruit is edible with a delicious, tangy flavor, though is not cultivated.	Low-growing	Alternate, odd-pinnately compound, 4–12 cm long; leaflets variable in number, margins crenate	Terminal and lateral racemes, sometimes forked; yellow, 1–1.2 cm across, lobes bend backwards; anthers yellow.	Berry, golden yellow or orange to deep red, round, 8–12 mm diameter, fleshy

Ground-covering plant	Acanthaceae	*Blechum pyramidatum*	Blechum	N	Arid lowlands and moist uplands	Sprawls on the ground, when growing alone, but will grow over other plants, in a mixed garden.	Low-growing	Opposite, simple; blade ovate to lanceolate, 2–7 cm long, often covered with fine hairs; margins mostly entire	Dense, tubular, purplish or white flowers that are 4-sided,`12–15 mm long. Flowers appear as 'a box within a box'.	Capsule, oblong, 6 mm long
Filler	Bromeliaceae	*Racinaea insularis*	Galápagos tillandsia, huicundu	E	Arid lowlands and moist uplands	Would be an excellent accent to cultivate on landscape trees. Non-parasitic ephiphytic.		Radiating cluster near base, simple; blade strap-shaped 13–45 cm long, often reddish near the base	Short spike at the top of a long stalk, short spikes 2–9 cm long, flowerless portion of stalk to 75 cm long, bracts present; white, 3 petals 5 mm long	Capsule, 2.3 cm long
Ground-covering plant	Fabaceae	*Crotalaria pumila*	Dwarf rattlebox	N	Arid lowlands and moist uplands	Smaller leaflets, flowers and fruits that *C. Incana*	2 m tall	Alternate, odd-pinnately compound; leaflets 3, blade elliptic, ovate, round, or obovate, 0.5–3.5 cm long	Racemes, opposite the leaves or terminal; yellow, 0.6–1.2 cm long; flower composed of 1 large petal, usually reddish to red-brown.	Legume, 1–2 cm long
Ground-covering plant	Mimosaceae	*Desmanthus virgatus*	Slender mimosa	N	Arid lowlands and moist uplands		2 m tall	Alternate, bipinnately compound, with 1–7 pairs of pinnae, leaflets 2–8 mm long; margins occasionally with hair	Axillary clusters; greenish white, 2.5–4 mm long	Legume, linear, 1.5–5.5 cm long, flattened
Seasonal theme plant	Solanaceae	*Solanum americanum* #	American black nightshade, glossy nightshade, hierba mora	N	Arid lowlands	Beautiful, dark fruit. May be grown in pots or landscapes as a seasonal theme plant. In the wild, grows in disturbed areas such as fields, roadsides and around buildings.	60 cm tall	Alternate, simple; blade ovate to ovate-lanceolate, 2–14 cm long	Lateral, umbel-like cymes; white, sometimes with purple, rotate, lobes 4–5 mm across, anthers yellow	Berry, dark purple to black, 5 mm in diameter

(continued)

Table 8.1 (continued)

Layer (after Rainer and West 2015)	Family	Scientific name	Common name(s)	Range	Habitat	Notes	Height	Leaves	Flowers	Fruit
Ground-covering plant	Verbenaceae	*Lippia strigulosa*	Purple lippia	N	Arid lowlands and moist uplands	Drought-resistant groundcover.	Low-growing	Opposite, simple; blade ovate, 14 cm long; surfaces hairy, margins serrate	Terminal racemes to 21 cm long; white, bluish, pinkish, or purplish; salverform with 5 lobes to 3.5 mm long, 2 lipped, upper lip 2 lobed, lower lip 3 lobed	Schizocarp, oblong, 3–4 mm long, separating into 2 sections when mature, enclosed in an inflated calyx with tiny hooked hairs
Ground-covering plant	Aizoaceae	*Sesuvium edmonstonei* #	Galápagos carpetweed	E	Coastal	Gorgeous groundcover that forms dense carpets over rocky and sandy terrain. During the cool season, takes on a bright orange-red color.	Low-growing	Opposite, simple, blade oblacceolate, 3 cm long, fleshy	Axillary, solitary, white, 5 petals to 6 mm long	Capsule, ovoid with a caplike lid
Ground-covering plant	Aizoaceae	*Sesuvium portulacastrum* ** #	Purslane	N	Coastal zone and arid lowlands	Excellent choice for sea-side plantings. Salt tolerant. Propagate via cuttings.	Low-growing	Fleshy, gray-green leaves that are greener in the wet season, and more reddish in the dry season.	Small, pink, star-shaped.	Capsule, ovoid with a caplike lid

Ground-covering plant	Mimosaceae	*Neptunia plena*	Sensitiva de agua, water dead and awake, water mimosa, water neptunia	N	Coastal zone and moist uplands	In the wild, this species is found in wet areas, usually with waterlogged soil.	2 m tall	Opposite, bipinnately compound; leaflets 10- many	Bright yellow, globose heads; petals united,	Legume, elongate and straight
Seasonal theme plant Ground-covering plant	Verbenaceae	*Phyla nodiflora*	Turkey tangle Frogfruit	N	Coastal zone, arid lowlands and moist uplands	Hardy, no-mow groundcover with small white and purple trumpet-shaped flowers. This plant is highly adaptable, including tolerating foot traffic and dogs. Requires less water than traditional turfgrass. Although it spreads via runners, it can be managed to stay within borders.	Low-growing	Opposite, simple; blade oblanceolate to obovate, 0.5–5 cm long, 0.2–3 cm wide' margin sharply serrate at apex; glabrous to pubescent	Axillary spikes; spikes purplish, cylindrical, 0.3–2.5 cm long; corolla mauve, pink, or white with yellow center, often white and purple in one inflorescence	Schizocarp, obovoid, 1.2 mm long; divided into 2 pyrenes

(continued)

Table 8.1 (continued)

Layer (after Rainer and West 2015)	Family	Scientific name	Common name(s)	Range	Habitat	Notes	Height	Leaves	Flowers	Fruit
Seasonal theme plant	Verbenaceae	*Verbena litoralis*	Vervain, seashore vervain	N	Moist uplands	Tall, strong square-sectioned plant with dividing candelabra-like stems and rough leaves. In the arid lowlands, could be included in container designs.	2 m tall	Simple, oblong leaves 3–10 cm long, somewhat hairy. Upper surface sometimes rough; margins coarsely serrate on upper two-thirds	Clusters of small purple, bluish, or pink tubular flowers, 2.5–3 mm across.	Schizocarp, oblong, 2 mm long, mostly enclosed in calyx, separating into 4 sections when mature
Ferns										
Filler	Pteridaceae	*Adiantum macrophyllum*	Helecho Cilantrillo, largeleaf maidenhair	N	Moist uplands	Fronds develop into a bright pink color	30–70 cm	Fronds	Not applicable	Not applicable
Filler	Dennstaedtiaceae	*Pteridium arachnoideum*	Tropical Brackenfern	N	Moist uplands	A beautiful fern with distinctive pinnule.	1.5–3 m tall	Fronds	Not applicable	Not applicable
Grasses and sedges										

Filler	Cyperaceae	*Cyperus anderssonii*	Andersson's sedge	E	Coastal zone, arid lowlands, and moist uplands	Single, upright stem that gives it a graceful look. Not to be confused with grasses, sedges have triangular and solid stems, compared to rounded and hollow stems of grasses. Adaptable to a variety of habitats.	70 cm tall	Alternate and basal, simple; blade linear to more than 70 cm long, 2–6 mm wide; lower portion purplish brown, completely surrounding stem	Terminal, umbel-like corymb, yellowish brown to reddish brown; calyx and corolla absent	Achene, dark brown, elliptical, 1 mm long, somewhat 3-sided
Filler	Poaceae	*Chloris virgata*	Feather fingergrass	N	Arid lowlands	Stems round, smooth, usually growing in clumps.	1 m tall	Blade narrow, to 25 cm long, basal sheath surrounding stem and ligule; surfaces smooth to somewhat rough, margins rough.	Spikelets on terminal, feathery, fingerlike racemes; racemes 5–6 cm long, yellowish brown;	Grain

(continued)

Table 8.1 (continued)

Layer (after Rainer and West 2015)	Family	Scientific name	Common name(s)	Range	Habitat	Notes	Height	Leaves			Flowers		Fruit
Filler	Poaceae	*Eragrostis cilianensis*	Gray lovegrass, Stinkgrass	N	Arid lowlands	Unpleasant odor when wet (hence its common name). Distinctive appearance, but may be best sited away from restaurant seating or high traffic areas.	40 cm tall	Blade narrow to 10 cm long, ligule and basal sheath. Both surfaces smooth to somewhat hairy. Ligule margin with fine hairs, leaf margins smooth.	Spikelets on terminal, spikelike panicles; panicles 5–10 cm long, yellowish brown; spikelet 2.5 mm wide, 6–12 florets, 2 glumes	Grain	Spikelets on terminal, spikelike panicles; panicles 5–10 cm long, yellowish brown; spikelet 2.5 mm wide, 6–12 florets, 2 glumes	Grain	Grain
Filler	Poaceae	*Eragrostis ciliaris*	Eragrostis	N	Arid lowlands	Distinct red heads that droop, providing a graceful counterpoint to more striking plants in a design.	40 cm tall	Alternate, simple; blade narrow to 10 cm long, ligule and basal sheath; both surfaces smooth to somewhat hairy, ligule margin with fine hairs, leaf margins smooth			Flattened, open, spikelets on terminal, spikelike panicles; panicles 5–10 cm long, yellowish brown; spikelet 2–3 mm long, 6–12 florets, 2 glumes		Grain
Filler	Poaceae	*Sporobolus indicus*		N	Coastal								Grain
Filler	Poaceae	*Sporobolus virginicus* **	Beach dropseed, Hierba de orilla, pasto de playa	N	Coastal	Excellent groundcover for coastal beaches. Tolerant of saltwater. Spreads by rhizomes to form large patches.	10–40 cm tall	Leaf sheaths are overlapping, and somewhat hairy.			Greenish-white to purplish spikelets.		Grain

Layer (after Rainer and West 2015)	Family	Scientific name	Common name(s)	Range	Habitat	Notes	Height	Leaves	Flowers	Fruit
Filler	Poaceae	*Distichlis spicata*	Saltgrass	N	Coastal	Excellent groundcover for coastal beaches. Tolerant of saltwater. Spreads by rhizomes to form large patches.	10–60 cm tall	Narrow, stiff, green.	Racemes 2–10 borne on central axis; solitary spikelets	Grain
Cacti										
Structural & Framework Plant	Cactaceae	*Jasminocereus thouarsii* **	Candelabra cactus, candelabro, cacto esbelto, cirio	E	Arid lowlands	Currently in wide use as an urban ornamental plant.	7 m tall	Represented by spines, white, yellow, reddish-brown, or black	Solitary, 5–9 cm long, 2–6 cm across, inner perianth parts yellow to yellow green, numerous stamens	Berries green to reddish purple, roundish to oblong.
Structural & Framework Plant	Cactaceae	*Opuntia megasperma* **	Prickly pear cactus, tuna	E	Arid lowlands and moist uplands	Currently in wide use as an urban ornamental plant.	6 m tall	3–11 mmm long, nonpersistent, usually not obvious. Spines 1–50 per cluster, yellow, brownish or black, 2.5–10 cm long; glochids few or absent	Solitary, 6–13 cm long, 6–10.5 cm across; inner perianth yellow to reddish yellow 2.5–5 cm long	Berry, green to yellowish green, top-shaped, 4–17 cm long, 2.7–5.5 across, fleshy, spines present

(continued)

Table 8.1 (continued)

Layer (after Rainer and West 2015)	Family	Scientific name	Common name(s)	Range	Habitat	Notes	Height	Leaves	Flowers	Fruit
Shrubs and trees										
Seasonal theme plant	Amaranthaceae	*Alternanthera filifolia* ** #	Thread-leafed chaff flower, Alternantera Hilo	E	Arid lowlands	Compact shrub whose shape can be maintained by pruning older branches.	1.5 m tall	Opposite, simple, blade linear to oblong, 2–5 cm long, 2–7 mm wide	Paper white flowers. 3–15 mm long, 3–5 mm across.	Utricle
Seasonal theme plant	Amaranthaceae	*Alternanthera echinocephala* ** #	Spiny-headed chaff flower, Alternantera blanca	N	Arid lowlands	Gorgeous shrub whose shape can be maintained with annual pruning.	2 m tall	Opposite, simple; blade lanceolate; 4–9 cm long	White, spiky flowers with exposed pollen that give the blooms a touch of yellow. Bracts and bracteoles are sometimes pink.	Utricle, 2 mm long
Structural & Framework Plant	Boraginaceae	*Cordia lutea* ** #	Yellow cordia, muyuyo	N	Arid lowlands	Produces a wonderfully sweet aroma and are a favorite source of nectar of *Phoebis sennae* and *Xylocopa darwini*. Wood is used for carvings that are sold locally	8 m tall	Alternate, simple; blade ovate to somewhat roundish, 4–10 cm long; upper surfaces rough, lower surface hairy; margins minutely crenate	Cymes; yellow, funnelform with 5–8 lobes, 2–4 cm across	Drupe, white, roundish, 8–12 mm across

Structural & Framework Plant	Burseraceae	*Bursera graveolens* **	Incense tree, Palo santo	N	Arid lowlands	Has aromatic resins with medicinal properties. Loses its leaves and trunk becomes a whitish gray during the cool season, becoming active when the rain returns.	3–12 m tall	Bright green, clustered at the branch tips. Pinnately compound, 10–28 cm long.	Axillary panicles 10–13 cm long; pale yellowish white	Drupe, brown when mature, roundish, 9–12 mm long
Structural & Framework Plant	Caesalpiniaceae	*Senna pistaciifolia* ** #	Flat-fruited senna	N	Arid lowlands	Similar to *Senna bicapsularis* and *Senna alata*. Attractive yellow flowers visited by *Xylocopa darwini*	1–2 m tall	Alternate, even-pinnately compound; leaflets 10–14, oval to oblong, longest leaves are 5–8 cm long	Axillary or terminal racemes, corolla yellow, 1.5–2 cm long	Legume, flattened and oblong; 7–11 cm long
Structural & Framework Plant	Caesalpiniaceae	*Parkinsonia aculeata* #	Jerusalem thorn, Palo Verde	N	Arid lowlands	With delicate drooping branches, is a common sight for tourists when visiting the archipelago	2–10 m tall	Alternate, bipinnately compound with 2–4 ribbonlike pinnae, 17–45 cm long; leaflets 20–30 pairs per pinna, oblong to elliptic, 3–10 mm long	Axillary racemes 6–20 cm long; yellow with one petal splotched reddish orange, 2–3 cm across	Legume, linear, 5–20 cm long, strongly constricted between the seeds
Structural & Framework Plant	Euphorbiaceae	*Chamaesyce viminea* **	Spurred chamaesyce	E	Arid lowlands	During cool season, the whole plant is reddish brown in color. Will lose leaves, but not die, without irrigation. Subject to scale insect infestations.	1.5 m	Opposite or clustered on short spur shoots; blades linear 1–4 cm long	Terminal, solitaru, green cyathia; each cyanthum possesses 4 black glands	Capsule

(continued)

Table 8.1 (continued)

Layer (after Rainer and West 2015)	Family	Scientific name	Common name(s)	Range	Habitat	Notes	Height	Leaves	Flowers	Fruit
Structural & Framework Plant	Euphorbiaceae	*Chamaesyce amplexicaulis*	Chamaesyce	E	Arid lowlands	During cool season, the whole plant is reddish brown in color. Will lose leaves, but not die, without irrigation. Subject to scale insect infestations.	60 cm tall	Opposite, simple, blade round, 3–11 cm long, 4–11 mm wide, cordate base wrapping around the stem	Terminal, solitaru, green cyathia; each cyanthum possesses 4 black glands with white to yellow green appendages resembling petals	Capsule, broadly ovate, 1.8 long, 2.3 mm wide
Structural & Framework Plant	Mimosaceae	*Vachellia macracantha* ** #	Acacia boton	E	Arid lowlands	Listed in Atkinson et al. (2017) as Acacia boton. Slow-growing shade tree.	8 m tall	Alternate, bipinnately compound with 5–30 pairs of pinnae; leaflets 14–24 pairs per pinna, oblong, 1–3 mm long and 0.5–1 mm wide; margins with fine hairs	Axillary, round clusters 8 mm across; yellow to orange; tubular with 5 lobes, 1–2 mm long	Legume, linear-oblong, 7–16 cm long, flattened
Seasonal theme plant	Scrophulariaceae	*Capraria biflora* **	Hairy capraria	N	Arid lowlands	Growth form is shrubby, but annually resprouts from the same roots.	1 m tall	Dark green, shiny, serrated, 3–11 cm long; glandular dotted on both surfaces.	Small (5–6 mm), greenish-white to pale greenish-yellow, star-shaped flowers. Anthers bluish	Capsule, brown, ovoid, 5–7 mm long, covered with glandular dots
Structural & Framework Plant	Verbenaceae	*Lantana peduncularis*	Galápagos lantana, supi-rosa	E	Arid lowlands	In the wild, can form dense thickets on lava and cinder fields. Loses its leaves in the dry season, but foliage can be maintained with irrigation.	2 m tall	Opposite, simple; blade ovate or ovate-lanceolate to elliptic; 7 cm long to 4 cm wide; surfaces hairy to velvety	Axillary heads; corolla white often with yellow throat; salverform with 5 unequal lobes, 8–9 mm long	Drupaceous, roundish

Seasonal theme plant	Malvaceae	*Gossypium darwinii* ** #	Darwin's cotton, Algodoncillo	E	Arid lowlands and moist uplands	Iconic plant of the Galápagos. Beautiful and distinct when in flower or with cotton-ball fruit.	1–3 m tall	Alternate, simple; blade usually with 3–5 lobes, occasionally unlobed and ovate, 15 cm long, 20 cm wide	Axillary, solitary; yellow and red purple at base of each petal- turns pink orange with age, 4–8 cm long petals	Capsule, ovoid, somewhat triangular in cross section, 1.8–2.8 cm long, usually 3 sections, covered with black pits
Seasonal theme plant	Rubiaceae	*Psychotria rufipes* ** #	White wild coffee, cafetillo	E	Arid lowlands and moist uplands	Resembles *Coffea arabica*, an introduced species, though the fruit and flowers of *P. Rufipes* are smaller.	1–3 m tall	Opposite, simple; blade broadly elliptic-ovate to obovate; 8–20 cm long, leathery and dark green, lower surface with red hairs	Terminal and axillary panicles; corolla white, tubular-salverform, 8–10 mm, covered in dense hairs	Berry-like, red, ellipsoid to roundish, 7–9 mm long, fleshy

(continued)

Table 8.1 (continued)

Layer (after Rainer and West 2015)	Family	Scientific name	Common name(s)	Range	Habitat	Notes	Height	Leaves	Flowers	Fruit
Structural & Framework Plant	Avicenniaceae	*Avicennia germinans* ** #	Black mangrove, mangle negro	N	Coastal zone	Flowers possess a pleasant odor, possess pneumatophores, grown well in waterlogged soil. Would be an excellent boardwalk planting, providing shade along a coastal promenade.	Varies from 2–25 m tall	Dark green, shiny leaves, 4.5–15 cm long; often with grayish fuzz on lower surface	White, creamy, or pale yellow campanulate flowers with 4 lobes.	Capsule, yellow then turns purplish black, 1.5–5 cm long, oblong
Structural & Framework Plant	Verbenaceae	*Lantana peduncularis*	Galápagos lantana, *supi-rosa*	E	Arid lowlands	*L. peduncularis* often forms dense thickets on lava and cinder fields	2 m tall	Opposite, simple; blade ovate or ovate-lanceolate to elliptic; 7 cm long to 4 cm wide; surfaces hairy to velvety	Axillary heads; corolla white often with yellow throat; salverform with 5 unequal lobes, 8–9 mm long	Drupaceous, roundish
Structural & Framework Plant	Bataceae	*Batis maritima* **	Saltwort	N	Coastal zone	Salt-tolerant ground cover. Tolerates partial shade. Minute flowers may be observed, though often overlooked	1.5 m tall	Opposite, simple; blade oblanceolate, 1–2 cm long, fleshy	Axillary conelike spikes; staminate spikes 3–8 mm long, greenish; calyx cup-shaped, greenish; pistillate spikes 5–20 mm long, greenish	Drupelike, those of each spike fused into a common structure

Structural & Framework Plant	Combretaceae	*Conocarpus erectus* ** #	Button magrove, mangle boton, jelí	N	Coastal zone	Gorgeous flowers and fruit	3–8 m tall	Alternate, simple; blade oblanceolate to narrowly elliptic, 2.5–9 cm long; somewhat leathery with a pair of raised glands at the base, small pits where the secondary veins meet the major vein on the lower surface; margins entire, sometimes waxy	Spherical heads borne in terminal panicles or racemes, calyx greenish	Drupelike, reddish brown, flattened and winged, 8–16 mm across
Structural & Framework Plant	Combretaceae	*Laguncularia racemosa* ** #	White mangrove, mangle Blanco	N	Coastal zone	Would be an excellent boardwalk planting, providing shade along a coastal promenade. Tolerates very saline conditions (90 ppt). Does not grow well, inland.	2–10 m tall	Thick, leathery leaves, 5.5–9 cm long. Margins smooth or slightly wavy.	Axillary or terminal paniculate spike, 7–8 cm long; corolla white, 2 mm long	Drupelike, gray-green, 1.5–2 cm long, somewhat flattened and winged with longitudinal ribs topped by a persistent calyx
Ground-covering plant	Goodeniaceae	*Scaevola plumieri* ** #	Inkberry, seagrape	N	Coastal zone	Considered rare in the Galápagos, planted on the mainland to reduce erosion along coastline, fruit adapted for oceanic dispersal.	1 m tall	Alternate, simple; blade obovate, 3–8 cm long	Cymes; corolla white, throat yellow and hairy	Drupe, green then dark purple or black, round, 1–2 cm long, possesses yellow flesh
Structural & Framework Plant	Rhizophoraceae	*Rhizophora mangle* #	Red mangrove, mangle rojo	N	Coastal zone	Another possible boardwalk planting, as an alee. Tolerates salt water to 60 ppt. Can form a shoreline thicket.	3–7 m tall	Opposite, simple; blades obovate to elliptic, 6.5–12.5 cm long, margins slightly rolled under	Axillary pairs, calyx greenish yellow, campanulate with 4 lobes 8–9 mm long, fleshy; corolla white to pale yellow, 4 petals 7–8 mm long	Berry, greenish brown, cone-shaped, 1.5–2.5 cm

(continued)

Table 8.1 (continued)

Layer (after Rainer and West 2015)	Family	Scientific name	Common name(s)	Range	Habitat	Notes	Height	Leaves	Flowers	Fruit
Structural & Framework Plant	Celastraceae	*Maytenus octogona* **	Maytenus, arrayancillo, rompe ollas, Galápagos leatherleaf	N	Coastal zone and arid lowlands	Slow-growing tree with fruit that splits open to reveal bright-red seeds. Very common on many islands.	8 m tall	Alternate, simple, coin-shaped leaves, 2–5 cm long. Thickened and leathery.	Solitary or in short axillary panicles, green; flowers 3–4 mm diameter	Green, somewhat roundish and 3-angled, 8–10 mm across, opening into 3 sections to reveal 1–3 bright-red seeds when ripe.
Groundcovers										
Ground-covering plant	Convolvulaceae	*Evolvulus convolvuloides* **	Purple evolvulus	N	Arid lowlands	Non-twining form, making it a manageable ornamental. Beautiful groundcover that is also adaptable to rock gardens.	50 cm long (sometimes longer)	Alternate, simple; blade broadly elliptic to somewhat round, 6–20 mm long, margins entire	Axillary, solitary; purplish or bluish, somewhat funnelform-rotate, 4 mm long, 7–10 mm across	Capsule, broadly obovoid to spherical, 3 mm long
Ground-covering plant	Euphorbiaceae	*Euphorbia thymifolia*	Red caustic-creeper	N	Arid lowlands	Drought tolerant groundcover. In the wild, can be found in disturbed sites in sandy soil. Valued for its medicinal properties.	Low-growing	Simple; blade ovate to 8 mm long and 4 mm wide; margins shallowly toothed; lower surface with long scattered hairs	Cyanthium; reddish and white	Capsule, 3-lobed, 1 mm wide, 1 mm long, pilose

Ground-covering plant	Boraginaceae	*Tiquilia galapagoa* **	Gray matplant	E	Arid lowlands	Spreading matlike herb	50 cm tall	Alternate, crowded clusters, simple; blade variously shaped but mostly ovate, 2–6 mm long; hairy	Dense clusters; 1 or 2 flowers open at a time; white, with a campanulate to funnelform or almost cylindrical tube, 3–4 mm long	Nutlet, dull black, <1 mm long, granular in appearance
Ground-covering plant	Convolvulaceae	*Ipomoea triloba* ** #	Morning-glory, Flor de la ma	N	Arid lowlands	In the wild, can be found growing in bright, open areas, as well as shaded forests. Currently in use as an urban landscape plant in some towns.	Stems to 4 m long	Alternate, simple; blade ovate to cordate, often 3-lobed, 4–10 cm long	Pink or purplish with a dark purple throat, occasionally white. Funnelform.	Capsule, somewhat spherical, 6–8 mm in diameter
Ground-covering plant	Convolvulaceae	*Evolvulus simplex* **	White evolvulus, Sogilla pequena	N	Arid lowlands					
Ground-covering plant	Portulacaceae	*Portulaca oleracea* ** #	Common purslane, verdolaga	N	Arid lowlands	Well suited for inclusion in container plant arrnagements, where it will spill over the sides.	Low-growing	Alternate or almost opposite, simple; blade wedge-shaped to obovate, 1–2.5 cm long, fleshy	Axillary clusters; yellow tepals, 3–8 mm long	Capsule, 4 mm long, opening by means of translucent lid
Ground-covering plant	Sapindaceae	*Cardiospermum halicacabum* **	Heartseed, Huevo frito	N	Arid lowlands	Occurs in disturbed areas in the wild. Is sometimes grown as an ornamental for its flowers, or to cover trellises. If not trained up a trellis, will cover the ground, but also other vegetation.	Climbing vine.	Compound, 3.5–14 cm long; 3 leaflets with 3 parts, margins serrate, dentate or slightly lobed	Axillary clusters; white, 4 petals 4.5 mm long, each with a yellow tipped appendaged	Capsule, brown when mature, roundish, 2.5–3.5 cm long, somewhat papery and inflated, 3-angled

(continued)

Table 8.1 (continued)

Layer (after Rainer and West 2015)	Family	Scientific name	Common name(s)	Range	Habitat	Notes	Height	Leaves	Flowers	Fruit
Ground-covering plant	Zygophyllaceae	*Kallstroemia adscendens*	No known common name	E	Arid lowlands	Similar to *Tribulus cistoides* (puncture vine), except that the fruits possess bumps, instead of spines.	Low-growing	Opposite, those of each pair unequal in size, pinnately compound, 2.5–8.5 cm long; leaflets usually 6–8 pairs, oblong to elliptic, largest ones 6–21 mm long, covered in tiny white hairs	Appearing axillary, solitary, 1.5–2.5 cm across; yellow, 5 petals 7–17 mm long	Schizocarp, grayish, hairy, to 1.5 cm wide, with 10 sections, each section with 2 large bumps (5–7 mm long) and 2 smaller bumps
Ground-covering plant	Solanaceae	*Exedeconus miersii* **	Galápagos shore petunia, Trompeta de orilla	E	Arid lowlands, coastal	Tolerant of salt water. Could be used as a ground cover in boardwalk plantings.	Low-growing	Hairy, light green, simple; blade ovate to ovate-reniform, 4–20 cm long; surfaces hairy, margins irregularly few toothed.	Axillary, solitary, sometimes appearing clustered; white with green veins, funnelform, 3–7 cm long, lobes 2–10 mm long	Capsule, elongate-ovate
Ground-covering plant	Boraginaceae	*Heliotropium curassavicum* ** #	Seaside heliotrope, cola de escorpion	N	Coastal zone	Beautiful scorpion's tale form of flowers. More prostrate. Could be groundcover.	Low-growing	Alternate, simple; blade elliptic or oblanceolate, 1.5–6 cm long; somewhat fleshy	Scorpioid spikes 10–15 cm long, often 2-cleft; white with yellow throat that may turn purple; funnelform 2 cm long	Nutlets, 2–2.5 mm across

Ground-covering plant	Commelinaceae	*Commelina diffusa* **	Dayflower, chiriyuyo	N	Moist uplands	Beautiful blue flower is highly delicate, lasting only half a day. Prefer somewhat shady areas, but can be found inhabiting open fields. Fed on by Galápagos tortoises	Low-growing perennial herb	Rooting occurs at nodes	Alternate, simple; petiole forming a sheath on the stem, sheath fringed with fine white hairs; blade ovate, 1.5–5 cm long, somewhat fleshy, margins entire	Terminal or axillary cymes; each cyme or pair of cymes with a large folded bract beneath; many flowers sterile; blue, occasionally white
Vining plants and climbers										
Seasonal theme plant	Passifloraceae	*Passiflora suberosa* ** #	Passion flower, uvilla, Grandilla de Monte	N	Moist uplands	Since it is a moist upland plant, could be a potted plant component, and maintained in arid lowlands with irrigation. As a vining plant, may be used as vertical interest up a post.	Vining	Alternate, simple; blade usually 3-lobed, occasionally unlobed; blade lanceolate to ovate, 3–7 cm long	Axillary, solitary or in pairs, 1–3 cm across; calyx greenish yellow; corona of numerous purple, yellow and white filaments	Berry, dark purple, oblong to somewhat ovoid to 1.5 cm long

(continued)

Table 8.1 (continued)

Layer (after Rainer and West 2015)	Family	Scientific name	Common name(s)	Range	Habitat	Notes	Height	Leaves	Flowers	Fruit
Seasonal theme plant	Passifloraceae	*Passiflora colinvauxii* ** #	Colinvaux's passion flower, granadilla silvestre	E	Moist uplands	Excellent choice for inclusion in a hanging basket arrangement. Honey-suckle scented passionflower. Pollinated by Darwin's carpenter bee. Withold fertilizer to avoid excessive leaf growth, and to push flowers.	Vining	Alternate, simple; blade usually 3-lobed, occasionally unlobed; blade lanceolate to ovate, 3–7 cm long	Axillary, solitary or in pairs, 1–3 cm across; calyx greenish yellow; corona of numerous purple, yellow and white filaments	Berry, dark purple, oblong to somewhat ovoid to 1.5 cm long
Seasonal theme plant	Menispermaceae	*Cissampelos pareira*	Hairy cissampelos	N	Arid lowlands and moist uplands	Bright red fruit of this climber can add visual, vertical interest.	5 m long	Alternate, simple; blade broadly ovate to roundish, 3–11 cm long, 2.5–10 cm wide; surfaces covered with soft hairs	Small, greenish.	Bright red drupe, 4–5 mm long, somewhat hairy.

Plants noted with double asterisks (**) are also recommended by Atkinson et al. (2017). Plants noted with a hashtag (#) are highly attractive to pollinators. Plants highlighted in light green are our favorite native and endemic plants for urban Galápagos landscape designs. Range includes native (N) and endemic (E) plants. This list drew heavily from Atkinson et al. (2017) and McMullen (1999), as well as observations posted to iNaturalist (2024)

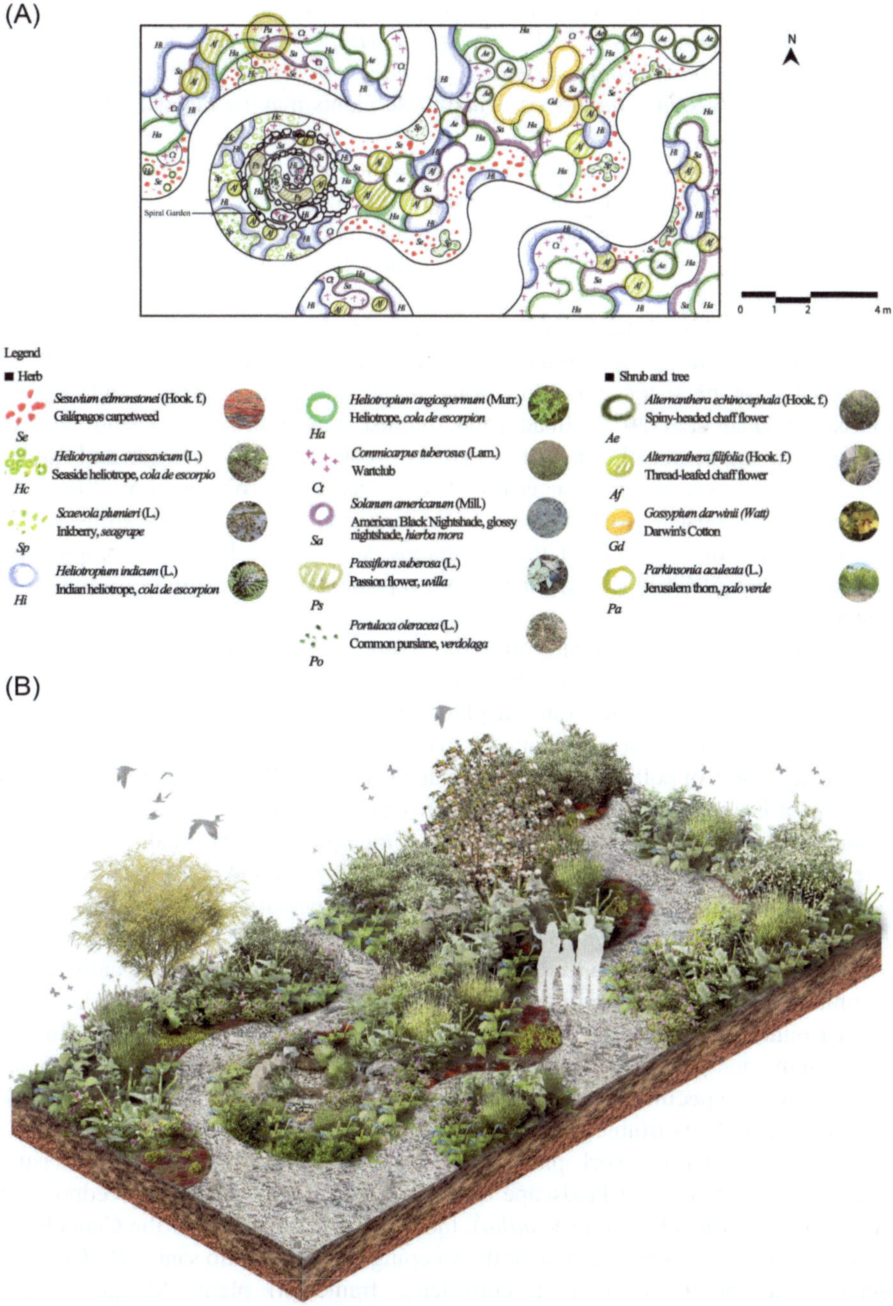

Fig. 8.1 Example garden design layout using plants native or endemic to the arid lowland regions of San Cristóbal, including (**a**) planting plan and (**b**) perspective rendering

Cristóbal landscape plantings. Finally, we removed plants with foul smelling or sticky fruit or plant parts (e.g. *Bidens pilosa*, *Hippomane mancinella*, *Ageratum conyzoides*, *Croton scouleri*, *Tribulus cistoides*, *Mentzelia aspera*, *Boerhaavia caribaea*, *Commicarpus tuberosus*) since these plants may be undesirable in high-traffic urban areas.

Our list of plants include annual and perennial herbaceous plants, cacti, ferns, grasses, groundcovers, and trees and shrubs, with notes on plant size. Plant categories are further organized by habitat (e.g. arid lowlands, coast, moist highlands) and include descriptions of ornamental plant features (e.g. flower color, leaf morphology, fruit structure).

Because of global interest in pollinator conservation, we highlight plants with high pollinator activity. Bees and tourism are indirectly related to one another, since bees enable the persistence and functional integrity of natural areas sought out by tourists (Adanan et al. 2017). In addition, interest in api-tourism (tourism related to beekeeping) has been gathering steam in recent years (Izquierdo-Gascón and Rubio-Gil 2023), as has a general concern for bee conservation (Wilson et al. 2017). As interest in bee conservation continues to grow worldwide (Hall and Martins 2020), the unique pollinator fauna of the Galápagos, such as the large and charismatic Darwin's Carpenter bee (McMullen 1989), represents an opportunity to draw in a new type of eco-tourist.

We thus identified pollinator-attracting plants using data from an iNaturalist project and two literature reviews (Chamorro et al. 2012; Traveset et al. 2013). From these three sources, we generated a plant-pollinator database of 775 records. Of these records, native plants accounted for one third of all plant species and over one half of all records of pollinator visits to Galápagos island plants. We cross-referenced these records with our list of Galápagos native and endemic plants with ornamental qualities (Table 8.1), to specifically call out forage plants for pollinators, including five with high levels of pollinator visitation: *Waltheria ovata* (Malvaceae), *Sesuvium portulacastrum* (Aizoaceae), *Cryptocarpus pyriformis* (Nyctaginaceae), *Heliotropium angiospermum* (Boraginaceae), and *Cordia lutea* (Boraginaceae).

We follow the suggestion of Rainer and West (2015), who note that a native plant community can be structured into four layers: (1) structural and framework plants, (2) seasonal theme plants, (3) ground-covering plants, and (4) filler species. In this way, ornamental plantings of native plants mimic natural communities in composition, but with a specific focus on design and aesthetics (Rainer and West 2015). We thus highlight plants from each of these design layers.

Structural and framework plants, distinct in shape and form, form the visual backbone of native plant landscape designs. The distinct spines and seedpods of Acacia botón (*Vachellia macracantha*), the striking architecture of the Candelabra cactus (*Jasminocerus thourarsii*), or the weeping canopy of Palo santo (*Parkinsonia aculeata*) are potential options to consider as framework plants. Mangrove trees (*Avicennia germinans*, *Laguncularia racemosa*, and *Rhizophora mangle*) can be planted in clusters or as an alee lining public boardwalk spaces, an approach that has been used in other landscape designs. These types of plantings provide shade and a natural experience for tourists, as well as the environmental benefits of shore protection and wildlife habitat (Fig. 8.2).

Fig. 8.2 Palo verde (*Parkinsonia aculeata*) blooming with flowers. (Photo credit Konstantinos Livadas)

Seasonal theme plants are those that visually dominate the planting for part of the growing year. The flowering phenology of many Galápagos plants has not been well-documented and/or collated. Thus, this could be a future area of research, to further the development and adoption of native plant landscape designs. Nonetheless, some plants to consider include Turkey Tangle Frogfruit (*Phyla nodiflora*), Darwin's Cotton (*Gossypium darwinii*), or Coffee senna (*Senna occidentalis*) (Fig. 8.3).

Fillers are usually fine-textured plants that add textural contrast or a colorful counterpoint. In monochromatic plantings, they demurely echo the beauty of structural and framework plants or seasonal theme plants. They add mass to landscape plantings, and provide a place for the eye to rest. Examples include Dombey's wood sorrel (*Oxalis dombeyi*), Galápagos jaegeria (*Jaegeria gracilis*), and False daisy (*Eclipta alba*).

Ground covers can be used to prevent erosion, or as a living mulch to suppress weeds. Some groundcovers, such as Galápagos shore petunia (*Exedeconus miersii*) can be used as a visual transition from land to sea, particularly when planted en masse along a shoreline. Some, such as Turkey Tangle Frogfruit (*Phyla nodiflora)* can be used as a low-water replacement for traditional turf.

Fig. 8.3 Flowers and foliage of a Turkey Tangle Frogfruit plant, Phyla nodiflora. (Photo credit weisschr)

Challenges of Native Plant in Urban Gardens and Greenspaces

The use of native plants in designed urban landscapes requires modifications to practices used in more traditional landscape designs. The water regime and soil in urban gardens and greenspaces are often very different than in the surrounding landscape. Gardens often exhibit excessive organic matter and fertility (Tresch et al. 2018; Nelson et al. 2022) compared to the native soils where native and endemic plants evolved. Within the islands, there has been some research for soil types, especially considering the islands of Santa Cruz, Santa Fe, and San Cristóbal. Five distinct soil zones have been identified using a case study of Santa Cruz Island (Stoops 2014). These include a dry coastal area characterized by reddish, smectite-bearing soils situated between basalt blocks from 100 to 120 meters above sea level (masl). Then transition zones with brown, smectite-bearing soils gradually merging the dry coastal area and higher andosolic soils at 100–180 masl. Higher zones above 180 masl then begin to exhibit deeper, brown soils with andosolic characteristics named "moist zones". These are divided into the lower from 180 to 300 masl, central from 300 to 400 masl, and upper from 400 to 500 masl (Stoops 2014).

Particularly, in San Cristóbal island, deeper, reddish, clayey soils are present due to the age of the island; but notably, no soils have been found on the younger islands with mostly sand and basalt (Fisk et al. 1982; Hoernle et al. 2000). All these areas display correlation to various vegetation types, with coastal smectite bearing soils being related to xerophytic vegetation, transition zones being related to dense forests of mesophytic and xerophytic associations, and high andosolic soils, or "moist regions", associated with *Scalesia* forests in their lower strata and with grassy or *Miconia* zones in higher strata (Reeder and Riechert 1975; Stoops 2014). Thus, efforts to amend and build soils in ways that are the norm for traditional garden soils may make it difficult for some native species to grow. For example, *Pectis subsquarrosa*, an endemic species to the Galápagos arid lowlands, grows best in fine gravel and with very little soil (Atkinson et al. 2017). This plant would be a good option for a dry riverbed or rock garden, but not in a traditional, soil-based landscape.

Fertilization and irrigation practices also need to be adjusted for native plant landscapes. Regular watering and fertilizing can provide a competitive advantage to plants that would be resource-limited in wildland habitats. The mere act of watering or fertilizing a garden bed, which is common practice for most gardeners and landscapers, can tilt the balance of a designed native plant community towards a single dominant species. Grouping plants from similar ecological zones (e.g. arid lowlands, transition zone, moist upland, highland zone) within a garden bed or landscape area can help alleviate this issue, because it segregates and groups plants by management needs. Plants from the arid lowlands (which represent the most urbanized region of the Galápagos) do not require much, if any, supplemental water or fertilization. Selecting low-maintenance plants from the arid lowlands can be a long-term design benefit because many gardens in the Galápagos receive little care from owners.

Particularly for plants native or endemic to the arid lowlands, many Galápagos native plants go dormant in the dry season, when they turn brown and lose flowers and leaves. This leads people to perceive these plants as dead, poorly cared for, or unattractive. However, this issue can be somewhat ameliorated with relatively small amounts of supplemental irrigation, which may extend the bloom season. Those designing, planting, and caring for Galápagos urban plantings could also take inspiration from renowned landscape designer, Piet Oudolf, who famously observed that 'brown is also a color' (McGrane 2008).

In terms of sustainability, in dry regions like Galápagos, where water conservation is paramount and soil nutrients are scarce, the strategic use of native plants in sustainable gardens becomes imperative. Galápagos lowland climate is characterized for its dry and cloudy season along one half of the year, with deciduous forests, semi-arid shrublands and lava deserts (Tallamy et al. 2010; Chamorro et al. 2012). Native species are naturally adapted to the local environment, requiring minimal irrigation once established and exhibiting resilience to drought conditions and lower mortality rates related to hydric stress (Bhatt 2013). Moreover, native plants have evolved to thrive in nutrient-poor soils, reducing the need for fertilizers, promoting overall ecosystem health and potentially renaturalizing and recovering derelict soils (Bretzel et al. 2009). By incorporating indigenous flora into garden designs, not only can water consumption be significantly reduced, but also the reliance on external inputs, fostering a harmonious coexistence with the environment while creating resilient landscapes and supporting native wildlife (Narango et al. 2017). And it is crucial to highlight that Galápagos human settlements have limited quantities of drinking water for human use (Grube et al. 2020). So, an effective way to reduce inefficient consumption and unessential consumption of water is to incentivize the use of native plant species as members of garden ornamentals and open green areas. And in relation to what was previously mentioned, native plants are ideal for San Cristóbal, where the gardens are more neglected, because they require less care from the owners.

Beyond the challenges imposed by the current gardening culture on the Galápagos, unique soils, and scarce water, one of the biggest challenges to promoting native plant landscapes on the islands is that they are not available for purchase, and they cannot be harvested from the Galápagos National Park. However, this also presents an opportunity to grow and support a native plant industry on the island, which could supply restoration efforts, as well as urban landscape needs. Local, native plant nurseries could sell to and supply multiple markets, including restoration, ornamental and urban plantings, and niche edible crops (e.g. *Physalis angulata*, Ground cherry).

In addition, there are numerous research gaps that could be filled, to inform the sustainable growth of a local native plant nursery industry. For example, very little is known about how many of these plants will perform when planted and maintained in urban settings. Basic data on plant performance, spread, and bloom phenology under a range of irrigation, sunlight, and maintenance conditions will help landscape designers better specify plants for particular circumstances, and can also help nursery growers know which plants have the best ornamental promise. Thus, the list of plants presented in Table 8.1 should be thought of as a starting point, where details on the successful cultivation and care of native plants can be filled in, as more data becomes available.

Summary and Conclusion

The use of native plants in Galápagos urban landscapes provides multiple benefits to diverse stakeholder groups. First, native plants, by definition, are not capable of becoming invasive (Iannone et al. 2020). They thus will not contribute to an invasive species problem that threatens local biodiversity and draws upon municipal resources for interception, surveillance, and management. Native plantings celebrate the unique natural heritage of the island and can echo or amplify the natural experience that high-value tourists are seeking from their Galápagos visit. They require less water than traditional landscape designs, which can help to conserve water for drinking and sanitation; a high priority need on the island (Grube et al. 2020). In addition, increased adoption of native plantings in the urban areas of Galápagos can create and grow new industries, including designers, nursery plant growers, and landscape maintenance companies that specialize in native plants.

References

Adanan NA, Rosmidi FH, Abdullah MT (2017) Establishing a link between bee conservation and ecotourism. Ecoutourism potentials in Malaysia. In: Mariapan M, Lin ELA, Isa SS, Karim MS, Hakeem KR (eds) Ecoutoursm potentials in Malaysia. Published by Faculty of Forestry Universiti Putra Malaysia, pp 48–52

Atkinson R (2008) Galápagos and its native gardens. In: Guzman I (ed) Annual Report. Charles Darwin Foundation, Puerto Ayora, Galápagos, Ecuador, pp 6–9

Atkinson R, Guézou A, Jaramillo Díaz P (2017) Siembrame en tu jardin – Kanpa sisapampapi tarpuway – plant me in your garden. Jardines nativos para la conservación de Galápagos

Beentje H, Jeffrey C, Hind DJN (2005) Compositae. Flora of Tropical East Africa

Bernays E, Graham M (1988) On the evolution of host specificity in phytophagous insects. Ecology 69:886–892

Bhatt A (2013) Propagation of potential native plants for urban landscape in Gulf countries. In Qatar foundation annual research forum (vol 2013, no. 1, pp EEP-06). Qatar: Bloomsbury Qatar Foundation Journals

Bretzel F, Pezzarossa B, Benvenuti S, Bravi A, Malorgio F (2009) Soil influence on the performance of 26 native herbaceous plants suitable for sustainable Mediterranean landscaping. Acta Oecol 35(5):657–663

Burghardt KT, Tallamy DW, Gregory Shriver W (2009) Impact of native plants on bird and butterfly biodiversity in suburban landscapes. Conserv Biol 23:219–224

Burghardt KT, Tallamy DW, Philips C, Shropshire KJ (2010) Non-native plants reduce abundance, richness, and host specialization in lepidopteran communities. Ecosphere 1(5):1–22

Burkart AE (1976) A monograph of the genus *Prosopis*. J Arnold Arboret 57:219–249

Carvache-Franco M, Carvache-Franco W, Manner-Baldeon F (2021) Market segmentation based on ecotourism motivations in marine protected areas and National Parks in the Galápagos Islands, Ecuador. J Coast Res 37:620–633

Chamorro S, Heleno R, Olesen JM, McMullen CK, Traveset A (2012) Pollination patterns and plant breeding systems in the Galápagos: a review. Ann Bot 110:1489–1501

Clem CS, Held DW (2015) Species richness of eruciform larvae associated with native and alien plants in the southeastern United States. J Insect Conserv 19:987–997

de Haan FJ, Quiroga Ferri D, Walsh SJ, Bettencourt LM (2019) Scales and transformative change: transitions in the Galápagos. In: Urban Galápagos: transition to sustainability in complex adaptive systems, pp 43–58

Diaz PJ (2019) Ecological gardens in the Galápagos: inspiring future generations. Charles Darwin Foundation News (July 11, 2019) https://www.darwinfoundation.org/en/news/all-news-stories/ecological-gardens-in-Galápagos-inspiring-future-generations/. Accessed 17 Mar 2024

Fisk MR, Bence AE, Schilling J-G (1982) Major element chemistry of Galápagos rift zone magmas and their phenocrysts. Earth Planet Sci Lett 61:171–189

Fontaine C, Thébault E, Dajoz I (2009) Are insect pollinators more generalist than insect herbivores? Proc R Soc B Biol Sci 276:3027–3033

Fukase J, Simons AM (2016) Increased pollinator activity in urban gardens with more native flora. Appl Ecol Environ Res 14:297–310

Gardener MR, Trueman M, Buddenhagen C, Heleno R, Jäger H, Atkinson R, Tye A (2013) A pragmatic approach to the management of plant invasions in Galápagos. In: Foxcroft LC, Pyšek P, Richardson DM, Genovesi P (eds) Plant invasions in protected areas: patterns, problems and challenges. Springer, Heidelberg, pp 349–374

Giefer M (2023) Land use and land cover change in the Galápagos: economic and natural drivers. In: Island ecosystems: challenges to sustainability. Springer International Publishing, Cham, pp 245–254

Grube AM, Stewart JR, Ochoa-Herrera V (2020) The challenge of achieving safely managed drinking water supply on San Cristóbal Island, Galápagos. Int J Hyg Environ Health 228:113547

Guézou A, Trueman M, Buddenhagen CE, Chamorro S, Guerrero AM, Pozo P, Atkinson R (2010) An extensive alien plant inventory from the inhabited areas of Galápagos. PLoS One 5(4):e10276

Hall DM, Martins DJ (2020) Human dimensions of insect pollinator conservation. Curr Opin Insect Sci 38:107–114

Hoernle K, Werner R, Morgan JP, Garbe-Schönberg D, Bryce J, Mrazek J (2000) Existence of complex spatial zonation in the Galápagos plume. Geology 28(5):435–438

Iannone BV, Carnevale S, Main MB, Hill JE, McConnell JB, Johnson SA, Enloe SF, Andreu M, Bell EC, Cuda JP, Baker SM (2020) Invasive species terminology: standardizing for stakeholder education. J Ext 58(3):Article 27

iNaturalist Community (2024) Observations of native plants from Galápagos Islands, Ecuador. Exported from https://www.inaturalist.org/projects/Galápagos-plants-with-ornamental-features-ii?tab=species on April 10, 2024

INaturalist Community (2025) White lead tree (*Leucaena leucocephala*) observations in the Galápagos Islands, Ecuador. https://www.inaturalist.org/observations?nelat=1.6818345&nelng=-89.2412769&subview=map&swlat=-1.4112351&swlng=-92.0089666&taxon_id=47447, search conducted on 6 June 2025

IUCN (2023) The IUCN red list of threatened species. Version 2023–1. https://www.iucnredlist.org. Accessed 27 May 2025

Izquierdo-Gascón M, Rubio-Gil Á (2023) Theoretical approach to Api-tourism routes as a paradigm of sustainable and regenerative rural development. J Apic Res 62:751–766

Jaenike J (1990) Host specialization in phytophagous insects. Annu Rev Ecol Syst 21:243–273

Kunze G (1847) Die farrnkräuter in kolorirten Abbildungen naturgetreu Erläutert und Beschrieben. E.F. Leipzig. (pp. 23). https://www.biodiversitylibrary.org/page/43085181#page/9/mode/1up

Lowe S, Browne M, Boudjelas S, De Poorter M (2000) 100 of the world's worst invasive alien species: a selection from the global invasive species database, vol 12. Invasive Species Specialist Group, Auckland

Majewska AA, Sims S, Wenger SJ, Davis AK, Altizer S (2018) Do characteristics of pollinator-friendly gardens predict the diversity, abundance, and reproduction of butterflies? Insect Conserv Diver 11:370–382

Mata L, Andersen AN, Morán-Ordóñez A, Hahs AK, Backstrom A, Ives CD, Bickel D, Duncan D, Palma E, Thomas F, Cranney K, Walker K, Shears I, Semeraro L, Malipatil M, Moir M, Plein M, Porch N, Vesk PA, Smith TR, Lynch Y (2021) Indigenous plants promote insect biodiversity in urban greenspaces. Ecol Appl 31:e02309

Matteson KC, Langellotto GA (2011) Small scale additions of native plants fail to increase beneficial insect richness in urban gardens. Insect Conserv Diver 4:89–98

Mazur C, Galush T, Moore R, Cotner S (2018) Primary motivations of tourists visiting Galápagos: do tourists visit the archipelago to learn about evolution? Evol: Educ Outreach 11:1–9

McGrane S (2008) A landscape in winter, dying heroically. New York Times

McMahan LR (2006) Understanding cultural reasons for the increase in both restoration efforts and gardening with native plants. Native Plants Journal 7:31–34

McMullen CK (1989) The Galápagos carpenter bee, just how important is it. Noticias de Galápagos 48:16–18

McMullen CK (1999) Flowering plants of the Galápagos. Comstock Publishing Associates

Mena CF, Quiroga D, Walsh SJ (2020) Threats to sustainability in the Galápagos Islands: a social–ecological perspective. In: The Elgar companion to geography, transdisciplinarity and sustainability. Edward Elgar Publishing, pp 342–358

Mickel JT, Beitel JM (1988) Pteridophyte Flora of Oaxaca, Mexico. Mem. New York Bot. Gard. 46: 1–580

Narango DL, Tallamy DW, Marra PP (2017) Native plants improve breeding and foraging habitat for an insectivorous bird. Biol Conserv 213:42–50

Nelson M, Mhuireach G, Langellotto GA (2022) Excess fertility in residential-scale urban agriculture soils in two western Oregon cities, USA. Urban Agric Reg Food Syst 7:e20027

O'Donnell PM (2004) Learning from world heritage: lessons from "International Preservation & Stewardship of Cultural & Ecological Landscapes of Global Significance. The 7th US/ICOMOS international symposium. The George Wright Forum, vol 21, pp 41–61
Pardee GL, Philpott SM (2014) Native plants are the bee's knees: local and landscape predictors of bee richness and abundance in backyard gardens. Urban Ecosyst 17:641–659
Rainer T, West C (2015) Planting in a post-wild world. Timber Press
Reeder WG, Riechert SE (1975) Vegetation change along an altitudinal gradient, Santa Cruz Island, Galápagos. Biotropica 7:162–175
Robertson HM (2022) A proposed tourism cap on the Galápagos Islands: beyond the wildlife. Wm Mary Env L Poly Rev 47:225
Salisbury A, Al-Beidh S, Armitage J, Bird S, Bostock H, Platoni A, Tatchell M, Thompson K, Perry J (2017) Enhancing gardens as habitats for plant-associated invertebrates: should we plant native or exotic species? Biodivers Conserv 26:2657–2673
Sardana K (2019) Tourists' willingness to pay for restoration of traditional agro-forest ecosystems providing biodiversity: evidence from India. Ecol Econ 159:362–372
Skolmen RG (n.d.) *Prosopis pallida* (Humb. & Bonpl. ex Willd.). USDA (United States Department of Agriculture). Retrieved November 7, 2023, from https://www.srs.fs.usda.gov/pubs/misc/ag_654/volume_2/prosopis/pallida.htm
Smith AR (1987) Euphorbiaceae. Flora of Tropical East Africa
Smith AR, Tejero-Díez D (2014) *Pleopeltis* (Polypodiaceae), a redefinition of the genus and nomenclatural novelties. Bot Sci 92:1
Stöcklin J (2009) Darwin and the plants of the Galápagos-Islands. Bauhinia:33–48
Stoops G (2014) Soils and Paleosoils of the Galápagos Islands: what we know and what we don't know, a meta-analysis. Pac Sci 68:1–17
Tallamy et al (2010) Can alien plants support generalist insect herbivores. Biol Invasions 12:2285–2292
Toral-Granda MV, Causton CE, Jäger H, Trueman M, Izurieta JC, Araujo E, Marilyn Cruz M, Zander KK, Izurieta A, Garnett ST (2017) Alien species pathways to the Galápagos Islands, Ecuador. PLoS One 12:e0184379
Traveset A, Heleno R, Chamorro S, Vargas P, McMullen CK, Castro-Urgal R, Nogales M, Herrera HW, Olesen JM (2013) Invaders of pollination networks in the Galápagos Islands: emergence of novel communities. Proc R Soc B Biol Sci 280(1758):20123040
Tresch S, Moretti M, Le Bayon R-C, Mader P, Zanetta A, Frey D, Stehle B, Kuhn A, Munyangabe A, Fliessbach A (2018) Urban soil quality assessment – a comprehensive case study dataset of urban garden soils. Front Environ Sci 6:136
Trueman M, Atkinson R, Guézou A, Wurm P (2010) Residence time and human-mediated propagule pressure at work in the alien flora of Galápagos. Biol Invasions 12:3949–3960
UNESCO 2023. Operational Guidelines for the Implementation of the World Heritage Convention, 29. https://whc.unesco.org/en/guidelines/
Verdcourt B (1992) Verbenaceae. In: Polhill RM (ed) Flora of tropical East Africa. Taylor & Francis Group, pp 1–156
Walsh SJ, Engie K, Page PH, Frizzelle BG (2019) Demographics of change: Modeling the transition of fishers to tourism in the Galápagos Islands. In: Urban Galápagos: transition to sustainability in complex adaptive systems, pp 61–83
Wilson JS, Forister ML, Carril OM (2017) Interest exceeds understanding in public support of bee conservation. Front Ecol Environ 15:460–466
Winkler IS, Mitter C (2008) The phylogenetic dimension of insect-plant interactions: a review of recent evidence. In: Specialization, speciation, and radiation: the evolutionary biology of herbivorous insects, pp 240–263

Chapter 9
Co-evolution Toward Harmony: Reimagining Sustainability Planning and Governance in Galapagos Landscapes Through Past Insights and Future Visions

Yizhao Yang and Jaime López-Andrade

Abstract This chapter examines the evolution of planning approaches in the Galápagos region from 1959 to the present, using an extensive literature review and analysis of regional-level plans to identify shifts in planning strategies. By assessing key aspects—primary driving forces, key actors and governance structures, and the scope and methods of planning—we identify four distinct phases: the Conservation-Centric period (1959–1979), focused on preserving biodiversity; the Conservation and Development phase (1980–1997), which balanced conservation with emerging developmental needs; the Integrated Conservation and Development era (1998–2014), emphasizing the integration of conservation and sustainable development goals; and the Synergistic Conservation and Development phase (2015–Present), which seeks to harmonize conservation and development through community-driven collaborative approaches. Building on this analysis, the chapter explores three critical areas—regional planning and development, placemaking for transformative effects, and evidence-driven collaborative decision-making—as essential components of a human-empowering policy framework to guide the Galápagos' urgent transition toward sustainability. These areas provide a foundation for addressing the region's complex challenges and fostering sustainable progress.

Y. Yang (✉)
School of Planning, Public Policy and Management, University of Oregon, Eugene, OR, USA
e-mail: yizhao@uoregon.edu

J. López-Andrade
Institute of Architecture and Urban Form (AFORU) and Instituto de Energía y Materiales, Universidad San Francisco de Quito (USFQ), Quito, Ecuador

J. Dunn Insua et al. (eds.), *Human Galapagos*, Social and Ecological Interactions in the Galapagos Islands, https://doi.org/10.1007/978-3-032-08800-0_9

Introduction

This chapter examines the evolution of planning approaches adopted to address the complex challenges facing the Galápagos region. We argue that since 1959 National Park Law, laying the foundation for the first policy intervention specifically designed for the archipelago, regional planning and policy formulation has been shaped by ongoing tensions between conservation and development, often reacting to emerging crises. Many of these crises have resulted from earlier policies' implementation or conflicts between policies pursuing competing objectives. Planning in the Galápagos has been largely driven by international pressure and national interests initially, rather than by local stakeholders within the region. As human settlements have expanded, their environmental impact has grown, yet these communities have also begun to exhibit adaptive growth dynamics within the Galápagos ecosystems (de Haan et al. 2019). Over time, policy shifts have become evident, particularly in the last 15 years, transitioning from an eco-centric framework to a more integrated approach—one that seeks to balance ecological conservation with local social needs and development.

The present moment marks a critical juncture for the Galápagos region to chart a path toward an active and sustainable transition. Such a transition, and the ongoing progress it requires, should be guided by policy approaches that avoid past mistakes, particularly the tendency to oscillate between extreme positions—either severely limiting human activity to protect the region's ecological integrity or promoting conventional economic growth at the expense of natural resources. Instead, a more viable and appropriate policy framework should recognize the interdependence of human and ecological systems, fostering mutually beneficial co-adaptation and co-evolution by embracing a socio-ecological planning approach.

Following the review of regional planning progression, this chapter then explores three critical areas—regional planning and development, placemaking for transformative effects, and evidence-driven collaborative decision-making—that underpin a more human-empowering policy framework to guide the Galápagos' urgent transition toward sustainability. We examine the theories and principles essential to successful actions in these areas, drawing inspiration from bioregionalism, the Sustainable Cities and Landscapes (SCL) model (Johnson and Ko 2022), and concepts linked to socio-ecological production landscapes and seascapes (SEPL&S) (Scheyvens et al. 2019). The discussion highlights the challenges and opportunities of implementing these principles in the Galápagos region, identifying existing positive trends and initiatives, such as participatory governance and community-led projects. Ultimately, we argue that a new planning phase for the Galapagos must not only respect the region's ecological history and protected status but also strive to transform its human settlements into vibrant, inclusive, and regenerative environments that embody the principles of bioregionalism and SEPL&S.

Research Methodology

Our approach to examining the evolution of Galápagos planning is informed by prior studies on conservation strategies (Valdivia et al. 2014), development models (González et al. 2008), and the driving forces behind regional transitions (de Haan et al. 2019). Additionally, we draw insights from Diego Quiroga's work, *Changing Views of the Galápagos*, which explores the changes in the perspectives of the islands and articulates the shift from an exclusively scientific focus to a hybrid approach that integrates conservation with community engagement (Quiroga 2013).

Our method of distinguishing planning and policy approaches from the late 1950s to the 2020s builds on an extensive review of literature on Galápagos environmental governance, combined with analysis of regional plans, policies, and legislative milestones, all situated within the global discourse on conservation and development during the same period. The literature review helps identify trends, shifts in priorities, prevailing challenges, key actors, and governance structures, while the plans, including formal national park management and regional development strategies, reveal the scope and methods of planning. The documents were retrieved from the digital repository of the National Institute of Advanced Studies of Ecuador, municipal archives, and the Official Gazette of the Ecuadorian government. Critical laws, beginning with the 1959 declaration of the archipelago as a state-controlled national park reserve, establish a dynamic legal framework that shapes the emergence of actors, spatial jurisdictions, and evolving stakeholder dynamics over time.

1. **The general objectives and primary driving forces** shaping Galápagos planning include international pressures, national interests, and local socioeconomic needs. Drawing on de Haan et al. (2019), we distinguish between external and intrinsic drivers of change and differentiate planning periods by considering the contrast between top-down, conservation-centric approaches and more integrated, socio-ecological perspectives. The latter increasingly emphasize local agency and sustainable coexistence between human and natural systems.
2. **The key actors involved in Galápagos policymaking and implementation**, which can be grouped based on their authority, expertise, and focus on community welfare. Drawing on Herremans and Ambrosie (2013), we identify three main groups shaping the Galápagos' planning processes. The key actors highlighted in each period are not an exhaustive list but rather those that represent significant shifts in planning and policy approaches.

    (a) **Formal Authorities**: Government entities, such as the Ecuadorian Ministry of Environment and the Galápagos National Park (GNP), wield formal regulatory power granted through legislation. The National Institute of Galápagos (INGALA), formerly responsible for coordinating and overseeing the management of the Galápagos Islands, held significant regulatory authority before being replaced by the Galápagos Government Council (CGREG).

These agencies establish and enforce conservation policies, manage resources, and regulate sustainable tourism, playing a crucial role in protecting the Galápagos ecosystems. Also included in this group are the Galápagos provincial government and various multisectoral commissions, which acted as representatives of the central government at different stages of policymaking and implementation.

(b) **Actors with Discursive Legitimacy:** These include international bureaucracies (e.g., WWF, UNESCO), conservation organizations (e.g., the Charles Darwin Foundation, CDF), scientific institutions (e.g., the Galápagos Science Center, the University of San Francisco de Quito), and non-profit organizations (e.g., Scalesia Foundation), which contribute specialized knowledge and recommendations to shape policies focused on ecological sustainability. While these actors hold significant influence due to their expertise, their role is primarily advisory rather than regulatory. As noted by Quiroga (2013), shifts in their influence and approaches reflect broader changes in the region's discourse, transitioning from science-based conservation to a more inclusive framework that incorporates local interests and needs over the past two decades.

(c) **Community-Oriented Stakeholders**: This group includes local governments, tourism operators and associations (e.g., the Galápagos Chamber of Tourism), as well as the Fishermen's and Farmers' Associations (e.g., Fisheries and Agricultural Cooperatives), which focus on the socioeconomic well-being of residents. Local governments, for instance, play a critical role in managing infrastructure, supporting sustainable tourism, and ensuring that public services align with conservation needs. They act as intermediaries, balancing community needs with conservation mandates.

In shaping Galápagos planning and implementation, each group of actors holds distinct values, interests, and priorities, even as they operate under the broader goal of sustainability. Over time, the influence of these groups has shifted. Initially, a top-down approach dominated, with formal authorities and international conservation organizations largely controlling planning and policy decisions. However, in the early 2000s, a co-management regime emerged, granting greater recognition to community-oriented stakeholders. This shift has not only elevated the role of local governments and economic stakeholders but has also encouraged conservation organizations and scientists to integrate community perspectives more directly into their work.

3. **Major regional plans and policies implemented**, reveal evolving approaches, shifting priorities, and changing responsibilities among key actors, particularly in balancing the dual dimensions of human activity and nature conservation. The main plans and policy instruments analyzed in our study include various editions of management plans for the national park and regional development plans for colonized areas or the entire region. These plans differ in three key aspects: (1) the level of integration, such as the extent to which conservation and development goals are combined; (2) the spatial scope, which ranges from focusing

solely on the protected national park to encompassing a broader region, including marine preserves; and (3) the evidence and input used for decision-making, which may rely on scientific research or information generated through inclusive, participatory processes.

Our approach employs primarily institutional-level analysis to identify the key phases in the evolution of Galápagos planning frameworks. In examining each phase, we link global trends in conservation and sustainable planning to the regional strategies developed for the Galápagos. While our analysis acknowledges the lived realities of local communities, both as contextual drivers of change and as subjects impacted by policy implementation, this chapter focuses on institutional shifts as the core factors shaping planning temporalities. We recognize this as a methodological limitation and emphasize that future research should explore the interaction between planning processes and communities. Furthermore, this research is largely descriptive, constructing a narrative of path dependency in Galápagos planning governance to establish a foundation for future strategic exploration. We do not investigate the specific mechanisms, such as stakeholder coalitions, that have propelled institutional change. Ultimately, this chapter presents one interpretive thread of governance and its role in Galápagos regional planning, offering a limited but necessary starting point for further research.

Regional Planning Approaches to Managing and Developing Galápagos

Pre-1959 Period

Prior to 1959, the starting point for this policy study, the Galápagos region had already drawn significant international attention for two key reasons. First, the publication of Charles Darwin's *On the Origin of Species* in 1859 highlighted the archipelago's unique biodiversity, establishing it as a global scientific treasure and a vital site for research in evolutionary biology, geology, oceanography, and ecology. This scientific value attracted the interest of international organizations like UNESCO and the WWF. Second, the islands' strategic location in the Pacific Ocean sparked geopolitical interest from nations such as the United States, England, and France in the nineteenth century, prompting the Ecuadorian government to initiate colonization efforts to assert control. The first Ecuadorian colonists arrived in the early 1830s, followed by groups led by national and foreign entrepreneurs contracted by the government in the 1850s and 1860s. However, a significant increase in the resident population did not occur until after World War II, facilitated largely by infrastructure development during the U.S. military presence from 1942 to 1946. Between 1941 and 1959, the population grew from 800 to approximately 2000 (Galápagos Conservancy n.d.-a).

Conservation efforts emerged in response to the challenge of balancing competing interests. The early initiatives, heavily influenced by the scientific community's focus on protecting the islands' unique ecosystems and biodiversity, laid the groundwork for evolving approaches over time. Early strategies focused on removing specimens from the Galápagos for protection. During the late nineteenth and early twentieth centuries, numerous expeditions collected specimens that were transported to laboratories and zoos worldwide. A notable example is the California Academy of Sciences expedition, which alone removed 75,000 biological specimens, including tortoises, reptiles, mammals, insects, and plants. By the early twentieth century, biologists began to recognize the importance of studying species within their natural habitats, leading to a shift toward in situ research (Quiroga 2013). This change in approach laid the foundation for subsequent legal and policy measures aimed at protecting the entire region for conservation purposes.

Laws protecting the Galápagos Islands were first enacted in the 1930s, designating the land as state property. In 1936, the Galápagos—then called the Archipelago of Colón—was declared a National Park and Flora and Fauna Reserve. Recognized internationally as one of the earliest efforts to prevent the destruction of Pacific islands, these laws prohibited hunting and the capture of birds and animals, allowing only limited extraction for scientific purposes. In 1937, all land was officially declared exclusive state property, with private ownership permitted only if legal claims could be proven for cultivated or occupied areas. While conservation was identified as a priority during this period, World War II delayed the formalization of related policies (Quiroga 2009). During the 1950s, UNESCO scientists visited the islands, emphasizing their ecological significance. This increasing international attention ultimately led to the creation of the Galápagos National Park in 1959. The designation excluded areas that had already been colonized, which constituted approximately 3% of the region. This ensured that the remaining 97% was protected and dedicated to ecological preservation.

Before delving into the discussion of the evolution of governance and planning in the Galápagos, we offer a concise overview of two key structural forces that facilitated policy implementation and transformation: land ownership and economic transitions. Informal land ownership, a legacy of early colonization, gradually transitioned into a regulated system in the 1960s and 1970s (Alomía Herrera et al. 2022). The year 1959 marked the start of formal, restrictive landownership delineation, limiting private ownership to just 3% of the land, coinciding with the 100th anniversary of the publication of "On the origin of Species" and the founding of the Charles Darwin Foundation (UNESCO World Heritage Centre n.d.). This restriction was subsequently upheld through strict migration policies, first introduced in the 1970s and further reinforced in the 1990s, which controlled the number of permanent residents eligible for land ownership. Economically, the region shifted from extractive industries, to agriculture, fishing and, eventually, tourism from the 1860s to the 1990s (Quiroga 2013). Recommendations to use tourism as a funding mechanism for Galápagos National Park conservation first emerged in the 1960s (Galápagos Conservancy n.d.-a). However, tourism only emerged as a dominant economic force in the 1990s, as fishing declined due to overexploitation and tighter regulations. Today, tourism drives approximately two-thirds of the local economy, with significant implications for the region's environmental sustainability and resilience. An emerging type of ecotourism that emphasizes benefiting both the environment and local population is being promoted by governments and embraced by local communities (Quiroga 2013).

1959- Present: The Changing Relationship Between Conservation and Development

The year 1959 marked a turning point in Galápagos planning and governance with the enactment of the National Park Law. This legislation established the foundation for conservation efforts by introducing dedicated policies, tools, and governance structures to protect the islands' unique ecosystems. Since then, ongoing and increasingly complex planning and policy changes have significantly shaped the region's environmental, as well as social and economic landscapes, whether intentionally or as unintended consequences.

We identify four distinct phases in Galápagos governance, each characterized by unique management and development approaches. These phases differ across three core dimensions analyzed through literature and policy review: (1) primary objectives and motivations, (2) key policymaking and implementing actors, and (3) major enacted plans and policies. Each phase's label reflects the evolving relationship between conservation and development priorities as evidenced in these dimensions. Table 9.1 summarizes this progression, detailing the driving forces, planning objectives, institutional actors, governance frameworks, and landmark policies that enabled each paradigm shift.

It is important to note that while we use clear labels and specific timeframes to define these phases and periods, policy practices often evolve gradually and reactively, influenced by historical decisions (path dependency). The roles of key actors in planning and policymaking were not immediately formalized; instead, their responsibilities and scope expanded and adapted over time. Therefore, the years marking each period should be seen as a framework to facilitate discussion rather than strict points of transition.

Conservation Centric Planning (1959–1979)

The period from 1959 to 1979 in the Galápagos was marked by a strong, almost singular emphasis on conserving the natural environment, driven largely by international recognition of the unique value of the region's ecosystem. This approach, reflecting the dominant international conservation model of the time, focused on designating critical areas for protection with strict limitations on human activities (Berkes et al. 1998). Supported by the global scientific community and driven by

Table 9.1 Major phases and periods of planning and governance in Galápagos (1959—Present)

Phase and period	Description of driving forces and planning goals	Legal foundation and landmark laws	Key actors and governance structure	Major regpional plans
Conservation Centric Planning (1959–1979)	During this phase, policies focused predominantly on national park protected areas, led by conservation organizations and implemented through environmental agencies. The central aim was ecosystem protection, with minimal consideration for human settlements or socioeconomic factors.	Declaration of the archipelago as a national park reserve under the exclusive domain of the State, 1959 Supreme decree for the creation of Galápagos Province, 1973	International science organizations (e.g., CDF, est. 1959) Ecuador National Government (est. 1973) UNESCO Galápagos National Park (GNP, est. 1959)	Management Plan of the Protected Area (1974) Conservation and Selective Development Plan (1975)
Conservation and Development (1980–1997)	Growing awareness of human impacts on ecosystems and conflicts between conservation and local needs spurred policies addressing regional socioeconomic development. However, these often remained disconnected from conservation efforts, following separate policy processes.		Galápagos National Park (GNP) Galápagos National Institute (INGALA, est. 1980) Multisectorial commissions	Management Plans for the Protected Area (1984, 1996) Regional Development Plans (1984, 1988) Management Plan for the Galápagos Marine Resources Reserve (1992, Not Implemented)

Integrated conservation and development planning via co-management (1998–2014)	Planning adopted a more integrated approach, recognizing that balancing human development with environmental sustainability was crucial for long-term conservation success. This led to policies that more effectively combined both dimensions.	Galápagos Special Law 1998 Ecuadoraian Constituion (2008)	Galápagos National Park (GNP) INGALA (dissolved 2008) Galápagos Governing Council (est. 2008) Participatory Management Board (PMB, dissolved 2015) Inter-institutional Management Authority (IMA, dissolved 2015) Co-management regime (involving both types of actors)	Galápagos Marine Reserve Management Plan (1999) Regional Plan for the Sustainable Development of Galápagos (2003) Galápagos National Park Management Plan (2005, 2014)
Synergistic Conservation and Development via Empowerment (2015-present)	A shift toward empowering local communities to affect and shape the region's transition. Policies increasingly seek to harmonize conservation and development goals by fostering stronger local engagement and collaboration between stakeholders, integrating human and ecological concerns in a more cohesive and sustainable way.	Organic Law on Galápagos the Especial Regime of Galápagos Province (2015) The Creation of Hermandad Marine Reserve (2022)	Galápagos National Park (GNP) Galápagos Governing Council Participatory Management Advisory Council (est. 2015)	Galápagos Sustainable Development and Land Use Plan (2015) Galápagos 2030 (2020) Management Plan for the Hermandad Marine Reserve (2023)

major international organizations, this "fortress model" of conservation enforced fixed boundaries and punitive regulations (Adams 2004). The plans and policies under this approach not only spatially restrict human activities but also establish a nature-human privilege order in their policy frameworks, often prioritizing ecological preservation over human presence and needs (West et al. 2006).

During this period, key actors from the scientific community and international organizations played a significant role in shaping policies for the Galápagos Islands. Among these were the Charles Darwin Foundation (CDF), UNESCO, and the International Union for Conservation of Nature (IUCN), along with the National Fisheries Institute, all of which were instrumental in advocating for conservation efforts and promoting scientific research in the region. In 1959, the Ecuadorian national government enacted the Law of Forestry and Conservation of Natural Areas and Wildlife, designating nearly the entire Galápagos land area as a protected national park, with the exception of established human settlements. This landmark legislation provided the legal framework for the creation of the Galápagos National Park Service (GNPS), which was formally established in 1968 as the primary institution responsible for managing the national park. Other key contributors to policy development during this period were primarily national-level government agencies, such as the National Board of Planning and Economic Coordination. Although the Galápagos Province was established in 1973, it did not emerge as a significant policy actor in spatial management and development policies until the 1990s (González et al. 2008). Throughout the 1960s and 1970s, conservation efforts were largely driven by national authorities and international collaborations, rather than by local governance (Heylings and Bravo 2007).

Policy efforts during this period focused on clarifying the responsibilities of administrative bodies such as the new Galápagos National Park (GNP), exploring tourism—particularly through floating hotels—as a tool to support both conservation and the national economy, and developing protocols for managing the protected area while allowing for selective development. It was not until 1979 that the Ecuadorian government, through an inter-ministerial agreement, formally established the definitive boundaries of the Galápagos National Park, designating 97% of the territory as protected parkland and 3% as colonized areas.

The first management plan for the protected area, developed in 1974 with input from UNESCO and the Charles Darwin Research Station, marked the beginning of systematic conservation management. It introduced basic guidelines and zoning for preservation, tourism, and scientific research (Heylings and Bravo 2007; Quiroga 2013). In 1978, UNESCO designated the Galápagos as the first World Natural Heritage Site, a milestone that underscored its global conservation significance. In 1975, the National Economic Planning and Coordination created the first development plan for the region, titled the Conservation and Selective Development Plan. This plan focused exclusively on colonized areas, which were later categorized based on land use into urban, rural, and infrastructure zones. That same year, a baseline study was conducted on marine communities across various sites, including inland waters. However, marine reserve areas were not established until the late 1980s, and formal management plans for these areas were not developed until the late 1990s. This gradual progression highlights the evolving understanding of the need to integrate marine ecosystems into the broader conservation framework.

Conservation and Development (1980–1997)

During the 1980s, international conservation practices shifted toward a more comprehensive ecosystem approach and inclusive model. This evolution emphasized the importance of addressing system-wide issues and fostering greater community involvement in conservation efforts (Berkes et al. 1998). This shift aligned with a broader, holistic approach to sustainable development, as articulated in the 1987 Brundtland Report, which integrated social and economic dimensions into sustainability discourse. One notable initiative reflecting this trend was that the Man and the Biosphere (MAB) Program, launched by UNESCO in 1971 to promote conservation and sustainable use of natural resources, designated Galápagos National Park as a Biosphere Reserve in 1984, recognizing the region's global significance and the need to balance nature conservation with human activities. These international influences contributed to an increasingly expansive perspective on ecosystem management in the Galápagos, extending efforts to encompass marine conservation, environmental education, and the promotion of volunteerism as a means to support nature conservation (CDF n.d.).

While external forces continued to influence planning practices, internal conditions in the Galápagos increasingly shaped decision-making, emphasizing the need to integrate human development with ecological sustainability. During this period, rapid population growth—primarily driven by migration from the mainland—became a major factor, with an annual growth rate of 6% in the 1990s (Bremner and Perez 2002). This growth was particularly concentrated in urban areas, alongside a significant economic shift from an agri-fishing-based economy to one dominated by tourism. This shift is reflected in the decline of the rural population in absolute number and in shares, which dropped from over 40% in the 1970s to below 20% within two decades (Epler 2007). These transformations created substantial internal pressures, underscoring the urgent need for strategic planning to balance human development with environmental protection.

International pressure to address ecological degradation in the Galápagos intensified in the 1990s, a series of events highlighted the growing human impact on the region's fragile ecosystems. A key driver of this pressure was the rapid expansion of the fishing industry, particularly the exploitation of sea cucumbers, which led to overfishing and sparked conflicts between local communities and conservationists (Quiroga 2013). These tensions reached a boiling point when protesters, dissatisfied with political decisions balancing development and conservation, occupied conservation facilities. The central issues revolved around conflicts over marine resources, such as overfishing, and the growing threat of invasive species (Snell 1996). The protests and other events of the late 1990s served as a catalyst for major legislative reforms, culminating in the landmark Special Law for the Galápagos (GSL) of 1998, which aimed to address these challenges through stricter regulations and a shift into more inclusive governance and management systems (González et al. 2008).

New stakeholders emerged in response to the increasingly polarized demands between preserving the Galápagos' unique ecosystems and supporting the local population's aspirations for socioeconomic growth. One pivotal institution that became directly involved in regional planning was the National Institute of Galápagos (INGALA), established in 1980, marking the beginning of this period. This organization was responsible for "coordinating regional planning, government funding (national, bilateral, and multilateral assistance), and technical assistance in Galápagos" (INGALA, 2011, cited in (Herremans and Ambrosie 2013). Its authority extended to infrastructure development for both human settlements and conservation efforts, as well as the regulation of social and economic growth in alignment with environmental priorities. INGALA aimed to ensure that development initiatives remained sustainable and compatible with the archipelago's ecological goals.

Other entities also emerged, including fishing cooperatives representing local economic interests, a broader range of scientific institutions, and various governmental and non-governmental organizations. Fishing cooperatives, which formed in the late 1980s and early 1990s, became a critical force in implementing conservation policies. Over time, these diverse actors contributed to the establishment of a co-management or polycentric governance system that would shape the region in subsequent years (Cáceres et al. 2022; Heylings and Bravo 2007).

While the recognition was clear that coordinated planning should involve key actors, including INGALA, GNP, and Galápagos governments at multiple levels, policies and legal frameworks for an efficient multi-agency governance model were lacking during this period (Herremans and Ambrosie 2013). Conflicts among these actors were evident, as they pursued connected yet still differing priorities and goals: the GNP focused on National Park management, INGALA on development, and local governments on community needs. Additionally, the prevailing attitude that viewed human settlements as the primary source of problems, placing pressure on ecosystem services, spreading invasive species, and causing habitat destruction, further fueled tensions among the key players responsible for guiding and managing the region's growth (Quiroga 2013).

Major planning policies continued to be developed separately for protected areas and human-use areas, with limited coordination between them in terms of objectives and strategies (Heyling and Cruz, 1998). For protected areas, the Galápagos National Park Management Plans, developed by GNP, included the 1984 and 1996 versions. Compared with the 1974 Plan, these plans contained more detailed information on biophysical aspects and area delineation, refined the zoning system with more detailed regulations and established key management strategies, such as licensing tour operators to ensure compliance with conservation guidelines, controlling visitor numbers at key sites to minimize ecological impact, deploying park rangers and monitoring systems to enforce regulations, and encouraging local participation in eco-tourism and guiding services to promote community involvement in conservation efforts (LaCondamine 2022).

The plans guiding and managing the development of the region, Regional Development Plan, were developed by INGALA in 1984 and again in 1988. These plans laid the foundation for policies aimed at coordinating development strategies in human-use zones, promoting eco-social development with a long-term vision toward the new millennium. It is worth noting that baseline data collected in the 1980s on population distribution, employment, infrastructure, and basic needs of residents, started informing planning. Social science studies received growing recognition, studying the connections between different activities: tourism, fishing, and migration (Bremner and Perez 2002).

Planning efforts and policies for marine protection and management included the declaration of the Galápagos Marine Resources Reserve in 1986, which was later formalized in 1988. The first management plan for the reserve was developed in 1992 by the Galápagos National Park (GNP) through a limited collaborative process. However, the plan was never effectively implemented due to several challenges, including insufficient funding, the absence of a clear governance structure, and the lack of enforcement mechanisms (Maestro et al. 2024). It was not until 1998, with the establishment of a new governance regime under the Special Law for Galápagos, that a more inclusive and participatory decision-making framework was institutionalized, allowing for the effective enforcement of marine conservation plans (González et al. 2008; Maestro et al. 2024).

Integrated Conservation and Development Planning via Co-management (1998–2014)

International conservation during this period evolved to become more interdisciplinary, integrating ecological, social, and policy dimensions, with growing recognition of indigenous traditional knowledge and the importance of knowledge co-production (Berkes 2009; Tengö et al. 2014). Critiques of the conventional Western science-based approach have emphasized the need for a stronger connection between social and ecological systems. They advocate for resilience and adaptive management practices that move beyond an interdisciplinary approach toward a truly transdisciplinary perspective (Folke 2004, 2006; Raufflet et al. 2000). Conservation practices became more inclusive and integrated, with ecosystem-based approaches supporting large-scale initiatives such as invasive species removal, native vegetation replanting, and integrated marine and terrestrial preservation (Burgiel and Muir 2010; Mačić et al. 2018). Reflecting these shifts, international frameworks and platforms provided guidelines and mechanisms for conservation and sustainable development goals. For example, the Millennium Ecosystem Assessment (2005) and the Convention on Biological Diversity (CBD) underscored the link between ecosystem health and human well-being (Amparo Alban 2020), while initiatives like the World Urban Forum and the Global Campaign for Sustainable Urbanization, led by UN-Habitat, emphasized knowledge exchange and stakeholder collaboration as key strategies for sustainable urban development (Elmqvist et al. 2019). These changes marked a transformation in conservation, making it more holistic and aligned with interconnected ecological and human development challenges.

In Galápagos, the growth of tourism-driven economic activities—such as commerce, hotels, and restaurants—boomed during this period, creating new socioeconomic dynamics unique to the region. The rapidly expanding land-based tourism resulted in growing concerns over the negative effects of development and its spillover into protected areas (Reale 2022). Urbanization accelerated, particularly on the most populated islands, where the urban population approached nearly 90% of the total population by the turn of the twenty-first century (González et al. 2008;). The rapid expansion of tourism infrastructure increased pressure on fragile ecosystems and intensified competition for limited resources, such as freshwater and land. Additionally, the introduction of invasive species, often inadvertently brought in by tourism and migration, posed a major threat to native flora and fauna (Quiroga 2013). Efforts to address these challenges frequently led to conflicts with local stakeholders, highlighting the complex interplay between economic development, conservation goals, and community needs in the Galápagos.

Major legislative actions during this period established a strong legal foundation for an ecosystem-based conservation approach and a co-management governance regime in the Galápagos. The Special Law for the Galápagos (GSL), enacted in 1998, formally created the Galápagos Marine Reserve (GMR), extending 40 nautical miles from the archipelago's baseline. The GSL also introduced a comprehensive legal and procedural framework to strengthen stakeholder collaboration and coordination in managing the islands' natural resources. The GSL implemented strict regulations on immigration, residency, and tourism, assigning INGALA greater responsibilities to "coordinate and manage the activities of different social, economic, cultural, and environmental stakeholders at the local, national, and international levels to ensure sustainable human development and conservation of the ecological systems of the Galápagos" (INGALA, 2011, cited in (Herremans and Ambrosie 2013, p. 5).

Additionally, the GSL led to the creation of two key management bodies: the Participatory Management Board (PMB) and the Inter-Institutional Management Authority (IMA). These institutions facilitated governance by bringing together representatives from local sectors, such as fishing cooperatives and the chamber of tourism, the international scientific community, including the Charles Darwin Foundation (CDF), and government agencies, such as the Galápagos National Park (GNP). For the first time, local communities and stakeholders were officially included in planning and decision-making processes, marking a significant shift toward inclusive governance (González et al. 2008). While introducing several measures for managing the protected areas, such as regulations regarding the introduction of alien species, the GSL also included provisions and incentives for environmental education and local participation in conservation, and immigration and residence control measures to control population growth (UNESCO World Heritage Centre 1998).

While progress was made, the co-management regime designed to protect ecological resources, particularly marine ecosystems, and promote sustainable development in human settlements has fallen short of expectations. This underperformance stems from several key challenges, including limited institutional capacity, insufficient scientific support for resource allocation and management decisions, and the absence of a unified long-term vision for regional sustainability that integrates both strategic planning and practical implementation mechanisms (Hoyman and McCall 2013). Additionally, the top-down approach to development created a disconnect between regional planning and local governance, leading to misalignment between regional policies and municipal-level implementation in the Galápagos (Cáceres et al. 2022).

The 2008 Ecuadorian Constitution reform, which recognized at the national legal framework the rights of nature and the importance of participatory process, initiated major governance reforms, which included the dissolution of INGALA, merging its functions with the Provincial Government to establish the Governing Council of Galápagos. The new Governing Council, composed of three national ministers and the elected mayors of the Galápagos, aimed to streamline administrative decision-making. These reforms also included the dismantling of the Participatory Management Board (PMB) and the Inter-Institutional Management Authority (IMA), replacing them with the Participatory Management Advisory Council in 2015, which was intended to support the Governing Council by ensuring stakeholder participation in decision-making processes related to conservation and development. These governance changes aimed to enhance coordination between national, regional, and local authorities, fostering a more integrated approach to conservation and sustainable development in the Galápagos, though their effectiveness remains subject to ongoing evaluation (Maestro et al. 2024).

This period saw efforts to address the lack of interdisciplinary teams, a clear research agenda, and institutional capacity—significant steps toward overcoming key barriers to effective planning and innovative co-management through evaluation research. There was a pressing need for participatory data collection methods that go beyond reliance on census data, a critical tool for implementing migration controls and categorizing residents. However, social science research remains insufficient for informing policymaking, revealing a clear bias toward biophysical studies and a lack of interdisciplinary approaches that examine the interconnected dynamics of society, economy, and ecology. Studies by (Santader et al. 2009) and (Castrejón et al. 2014) point out that many socio-ecological processes—essential for sustainable planning and management, including marine management which requires both marine and management sciences—are still understudied. National strategic goals have since emphasized setting research priorities based on Galápagos needs, fostering integrated long-term socio-ecological research, building bridges with local institutions and stakeholders, and forming interdisciplinary research groups with Ecuadorian scientists to establish a robust Galápagos scientific community (Castrejón et al. 2014).

While new versions of the plans continued to be developed separately for protected areas, the marine reserve, and human-use territories, efforts to integrate these spatial plans into a unified framework with broader stakeholder participation gained momentum. The 1999 Galápagos Marine Reserve Management Plan governed one of the world's largest marine protected areas, introducing a zoning system to regulate fishing, ecotourism, and scientific research. The 2005 Galápagos National Park Management Plan focused on the terrestrial protected area. Yet, in 2014, the "Management Plan for the Protected Areas of Galápagos for Good Living" was introduced, unifying management strategies for both the Galápagos National Park and the Galápagos Marine Reserve. This plan strengthened conservation efforts by ensuring the protection of marine ecosystems as critical extensions of terrestrial biodiversity, reinforcing an integrated approach to sustainable management. The 2003 Regional Development Plan, still developed by INGALA, began incorporating concepts such as carrying capacity and socio-ecosystem frameworks, recognizing the interconnections between human activities and natural systems. The plan also introduced scientific methodologies to regulate visitor interactions with protected areas, ensuring tourism remained within sustainable ecological limits.

However, the problem of disconnection persisted, with a lack of coordination between the development of human settlements and the management and conservation of protected park areas. With rapidly rising, uncontrolled tourism driving urban expansion and infrastructure development, illegal construction and environmental degradation intensified, pushing the region's ecosystems toward serious deterioration. In 2007, the Galápagos faced the risk of losing its UNESCO World Heritage designation, which highlighted the urgent need for stricter conservation measures, improved governance, and sustainable planning. This crisis underscored the necessity of an integrated, long-term strategy that effectively harmonizes development with conservation, strengthens institutional capacity, enhances regulatory enforcement, and fosters a participatory approach to decision-making. Significant progress in these areas has been observed in the following years, especially since 2015.

Synergistic Conservation and Development via Empowerment (2015–Present)

Since 2015, international frameworks, particularly the 17 Sustainable Development Goals (SDGs), have significantly shaped Galápagos planning approaches (United Nations n.d.). The SDGs provide a global framework with concrete targets, guiding policy efforts and shaping the region's sustainable future. Goals such as SDG 14 (Life Below Water), SDG 15 (Life on Land), and SDG 12 (Responsible Consumption and Production) have influenced initiatives in sustainable fisheries, biodiversity conservation, and responsible tourism. Beyond the SDGs, other international frameworks have also influenced policies on local economic activities and sustainable agriculture, such as the Global Sustainable Tourism Council (GSTC) criteria (Galápagos Conservancy n.d.-b) and the Food and Agriculture Organization's (FAO) Climate-Smart Agriculture framework (FAO n.d.). Additionally, the UN Guide for Urban and Territorial Planning reinforces participatory planning as a core principle, encouraging inclusive decision-making, sustainable urban development, and balanced territorial management (UN-Habitat 2015). Collectively, these frameworks have deeply influenced planning processes in the Galápagos.

During this ongoing period, human settlements in the Galápagos have expanded significantly, reaching a scale at which local socioeconomic dynamics can no longer be overlooked in regional policy and planning. This growth has fostered unique settlement dynamics that now follow a path-dependent, evolutionary trajectory, intersecting with top-down policy initiatives. Local communities have increasingly demonstrated a remarkable capacity for innovation and adaptation, exemplified by efforts in local economic diversification, community-driven tourism, and artisanal fisheries (de Haan et al. 2019). These innovations have been crucial not only for charting sustainable development paths but also for preserving the region's ecological integrity by promoting resilience and local stewardship. Consequently, human settlements have evolved from passive recipients of policy into active agents, significantly influencing and shaping the socio-ecological landscape of the Galápagos.

The beginning of this period is marked by the enactment of the 2015 Organic Law for the Special Regime of the Galápagos Province, an expansion of the 1998 Galápagos Special Law (GSL) that introduced significant institutional reforms. Notably, it repealed the Participatory Management Board (PMB) and the Inter-institutional Management Authority (IMA), replacing them with the Participatory Management Advisory Council in 2015. This council serves an advisory role to the Galápagos Special Regime Governing Council, which was established in 2008. The reform aimed to create a cohesive, participatory, and forward-thinking framework for the entire archipelago, effectively integrating marine and terrestrial planning while providing a long-term strategic vision for sustainable development (Maestro et al. 2024).

The impact of an integrated governance structure is evident in the Galápagos Sustainable Development and Land Use Plan (2015–2020), approved by the Galápagos Special Regime Governing Council in 2015. This plan is the first to encompass the entire region within its planning scope. It emphasizes integrated governance, inclusive stakeholder engagement, and context-sensitive placemaking strategies, adopting a holistic, transdisciplinary approach that applies evidence-based methods to manage the relationship between human and natural systems. The Galápagos 2030 Plan, created in 2020 during the COVID-19 pandemic, incorporated resident participation and marked a significant policy shift. Introduced by the forward 'Investing in People is Investing in Conservation,' this plan expanded beyond conservation, instead managing the archipelago as a unified system, empowering local communities to actively shape the region's future.

To implement the integrated regional planning approach, the 2016–2020 zoning system was developed, encompassing both terrestrial and marine areas, integrating the entire archipelago under a unified zoning framework containing five types of zones (see Fig. 9.1). Human-use zones (yellow) encompass inhabited areas, urban, rural, and transitional, where human activities like living, working, and tourism are managed to align with environmental stewardship. Sustainable Use zones (green) make up most of the surrounding marine area, permitting regulated fishing and other sustainable practices to support local livelihoods while safeguarding marine ecosystems. Conservation zones (brown) are high-priority areas with restricted access, prohibiting extractive activities to protect critical habitats and vulnerable species. Intangible zones (orange) are the most strictly protected, often entirely off-limits to the public and

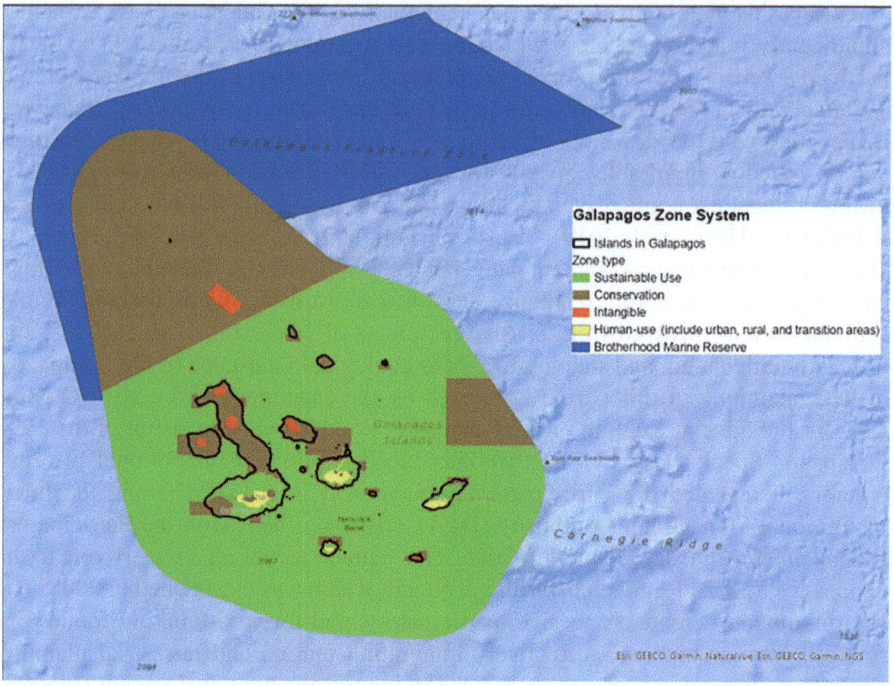

Fig. 9.1 2017 zoning system for the Galápagos

researchers, preserving pristine ecosystems in their natural state. Finally, the Hermandad (Brotherhood) Marine Reserve (blue), a newer addition beyond the main archipelago, serves as a symbolic and ecological link to regional biodiversity corridors and is designated a strict "no-take zone" to prevent any extractive use.

The management and protection of the Galápagos Marine Reserve (GMR) continue to expand. In 2022, the Hermandad Marine Reserve was added, extending the GMR by 60,000 square kilometers to protect critical migratory pathways and establish a marine corridor linking the Galápagos Islands with Cocos Island in Costa Rica (see Fig. 9.1). In 2023, the Galápagos National Park (GNP) developed a management plan for this new area, outlining its vision, objectives, and strategies for conservation and sustainable resource use. The plan includes zoning regulations, clearly defining areas where extractive activities are prohibited and areas where responsible fishing is allowed, and establishes programs for monitoring, enforcement, and local community engagement to ensure effective management and stakeholder participation.

Over the past 70 years, the Galápagos Islands have been a global leader in pioneering conservation strategies, marked by evolving governance structures, planning frameworks, and implementation methods. Key initiatives, such as the establishment of a National Park, large-scale eradication of invasive species, the creation of extensive marine reserves, and the use of zoning strategies within the archipelago, were among the first of their kind at that time. These efforts have underscored the importance of continuously adapting to emerging threats and pressures. As we transition to the next section, the focus shifts from examining the role of human actors in conservation to designing strategies and processes that foster activities creating synergies between developed and protected areas.

Achieving Human-Nature Co-evolution via Creating Socio-ecological Systems Approaches

Sustainability Planning for Ecologically Sensitive and Insular Regions

Our analysis of Galápagos planning approaches demonstrates the convergence of global trends and local dynamics, though both spheres continue to evolve in understanding the consequences of past decisions and necessary corrections. This research contributes to the literature on sustainability governance for fragile island ecosystems. Specifically, we demonstrate that the Galápagos planning for sustainability aligns with contemporary scholarship on island sustainability governance through three key insights: (1) the necessity of political-economic analysis to examine conservation-development conflicts shaped by multi-scalar governance and institutional influences; (2) the importance of decentralized, collaborative and adaptive governance models for strengthening resilience; and (3) the crucial role of socio-ecological systems approaches in fostering adaptability. These findings also highlight participatory processes as fundamental while recognizing the distinct challenges posed by islands' geographical constraints (Connell 2018; Kvan and Karakiewicz 2019; Raman et al. 2024; Uchiyama et al. 2023).

Progress now depends on translating scientific knowledge into strategies tailored to the unique conditions of the Galápagos Island system. Scholarly work has increasingly focused on this challenge, supported by growing research both about and within the archipelago. The Galápagos Science Center—established in 2011 through a partnership between USFQ and UNC and based in San Cristóbal—exemplifies efforts to advance such interdisciplinary research, which has expanded significantly in both social and economic dimensions. A prominent example is the 2019 volume Urban Galapagos part of a Springer-published series, which applies the complex adaptive systems (CAS) framework to analyze the Galápagos (Kvan and Karakiewicz 2019). This work demonstrates how human and natural systems function as interconnected systems while highlighting society's adaptive capacity to shape its own evolutionary trajectory.

In developing sustainability planning approaches for the Galápagos, we focus particularly on the key theme in the literature: the socio-ecological systems (SES) framework, which facilitates the co-evolution of human and natural systems for long-term resilience (Berkes et al. 1998; Nel et al. 2021). This approach recognizes the dual reality of human-environment relationships: while human activities undeniably impact ecosystems, they also contain the capacity to actively restore and protect nature. This understanding reflects a fundamental paradigm shift from 'managing nature through controlling people' to 'investing in people as a conservation strategy' (Raymond et al. 2022), representing a transition from restrictive conservation policies to empowerment-based models that transform traditional environmental governance.

Effective strategies must be grounded in continuously updated knowledge and profound engagement with local communities, recognizing both the challenges and opportunities inherent in sustainable development. Our 70-year analysis of Galápagos planning reveals parallel progress in two interconnected domains. Sustainability science has evolved from narrow ecological studies to holistic examinations of human-nature interdependence, incorporating socio-economic dimensions within complex adaptive systems. Concurrently, policy development has also advanced through iterative learning toward an integrated SES framework that harmonizes governance, spatial planning, and regional development. This dual progression underscores a transformative alignment—where contemporary strategies simultaneously preserve ecological integrity and enhance community well-being, achieving a balance between conservation goals and sustainable livelihoods.

Toward a Socio-Ecological Systems Framework for Action in the Galápagos

Within the SES framework, we identify three key action areas critical for advancing the Galápagos' paradigm shift toward sustainable coexistence: (1) bioregional planning methodologies, (2) transformative placemaking strategies, and (3) evidence-based adaptive collaboration. These interconnected components constitute an investment-oriented model that fosters reciprocal relationships between human and natural systems. By creating productive socio-ecological landscapes where both systems co-evolve, they establish the necessary conditions for sustainable transformation. The following discussion presents the theoretical foundations, operational principles, and implementation pathways tailored to the Galápagos context, emphasizing community-driven knowledge co-production as the essential foundation for developing context-appropriate sustainability policies.

Bioregionalism Informed Regional Planning and Development

Bioregionalism informed regional planning emphasizes aligning human settlements with the natural ecological boundaries that shape them—a concept encapsulated by Sir Patrick Geddes, who famously asserted, *"It takes the whole region to make the city"* (Geddes 1898, 1915). Geddes's pioneering work laid the intellectual foundation for integrating ecological and human systems, advocating for methods like the bio-regional survey to understand and manage the intricate interactions within a region. Building on this legacy, modern regional planning has expanded its analytical and methodological toolkit to include design with nature principles (McHarg 1995), which align human development with natural processes, and ecosystem services computing (Marques et al. 2022), which quantifies the benefits provided by ecosystems (e.g., clean water, carbon sequestration, pollination) to inform decision-making. Advancements in geospatial technologies (e.g., GIS and remote sensing) further enable planners to map and analyze ecological and socio-economic data at regional scales with unprecedented precision (Thompson et al. 2024).

These perspectives and methods inform the framework of Sustainable Cities and Landscapes (SCL), collectively developed by a network of researchers, educators, and practitioners affiliated with the APRU SCL Hub at the University of Oregon (Johnson and Ko 2022). The SCL framework builds on the intellectual foundations of landscape urbanism (Waldheim 2006, 2016) and ecological urbanism (Mostafavi et al. 2016), while also contributing to the discourse on bioregionalism-informed regional planning by drawing insights from global practices. It highlights the value of learning-by-doing and the development of information and knowledge infrastructures to support policy and technological transfer (Yang et al. 2022), ensuring that regional planning is both context-specific and globally informed.

Principles guiding the development of regional planning and development strategies have proliferated. Some examples include frameworks such as the UN Habitat 10 Action Principles for Urban-Rural Linkages (UN-Habitat 2019), which emphasize the integration of urban and rural areas to promote balanced development, and the Smart Growth Principles from the US (Smart Growth America n.d.), which advocate for compact, walkable, and transit-oriented communities. Additionally, the Garden City Principles from the UK (Town and Country Planning Association n.d.), when applied at a regional scale, promote the creation of self-sustaining communities surrounded by greenbelts, blending urban and rural elements. These frameworks, largely emerging from Western, developed countries, share a common focus on addressing the necessary issues and approaches to create and manage Socio-Ecological Production Landscapes and Seascapes (SEPL&S), where human livelihoods and ecological stewardship are mutually reinforcing (see Satoyama Initiative 2010). The common thread among these principles is their aim to foster sustainable development that respects ecological limits while enhancing the quality of life for communities. They achieve this by devising actions and processes tailored to the unique characteristics of a region—its ecosystems, resources, and cultural heritage.

Applying a bioregional planning approach to the Galápagos means grounding its principles and methods in the unique context of the archipelago, which calls for both action-oriented research and policy experiments. For instance, the natural boundaries and dynamic processes of the region's ecosystems necessitate an evidence-based and critical re-examination of existing policy boundaries, such as land use zones delineating urban, rural, and agricultural functions, to realign them with ecological processes like watersheds, climate zones, geological formations, soil types, and native vegetation. Considering the ubiquitous volcanic landscape of the Galápagos in the planning process can aid in the proper management of permeable and impermeable soil covers, protect underground waters from pollutants, and restore natural drainage systems—crucial mechanisms for mitigating the adverse effects of climatic phenomena. Research is also needed to explore strategies that integrate ecological health with cultural, social, and economic well-being, such as projects focused on restoring native species in urban areas or encouraging agricultural practices that control invasive species.

More importantly, a regional approach demands research and investment in developing local food systems, renewable energy solutions, and circular economic practices to reduce dependence on distant or non-renewable resources. The four inhabited islands—Santa Cruz, San Cristobal, Isabela, and Floreana—each have the potential to achieve this goal. For example, Santa Cruz and San Cristobal, with their larger populations and infrastructure, could lead efforts to connect urban areas with inland resources, such as agricultural lands and freshwater systems, while fostering sustainable tourism and fisheries. Establishing stronger connections among the islands and with the larger region is also critical to support the circulation and flow of goods and resources, reducing reliance on imports from the mainland and foreign countries.

Placemaking for Multifunctional and Transformative Outcomes

Integrating detailed and context-responsive placemaking strategies into the bioregional planning framework is essential for realizing Socio-Ecological Production Landscapes and Seascapes (SEPL&S)—a mosaic where human development and natural habitats coexist and mutually reinforce one another. These strategies should promote a creative and synergistic integration of built and natural elements across multiple scales, from individual buildings and sites to neighborhoods and entire cities. Drawing from foundational theories such as Kevin Lynch's *"The Image of the City,"* which emphasizes legibility, wayfinding, and the mental mapping of urban spaces (Lynch 2008), and Ian McHarg's *"Design with Nature,"* which advocates for ecological planning rooted in the understanding of natural systems (McHarg 1995), these approaches can create balanced and functional environments. Additionally, Jan Gehl's theory of *"Life Between Buildings"* underscores the importance of walkability, public spaces, and human-centered design in fostering vibrant, socially interactive communities while maintaining ecological integrity (Gehl 2013, 2015).

Synthesizing these theories leads to three core principles that can guide placemaking to deliver ecologically regenerative and socially inclusive outcomes. The first principle is to embrace multifunctionality in the design of human settlements. This means creating physical spaces that serve dual purposes—meeting human needs while supporting ecological functions. For example, public spaces can be designed to host community gatherings while also managing stormwater through green infrastructure. Managing parks and recreational spaces with native and endemic vegetation, along with sustainable urban drainage solutions that incorporate permeable natural coverage into park design, is essential. Similarly, green buildings can reduce environmental impacts while generating renewable energy through solar panels. Urban rooftop areas in the Galápagos offer thousands of square meters of potential for energy production from solar radiation, providing ample space to install sufficient capacity to satisfy all of Galápagos society's energy needs and thereby reducing the need for extensive solar gardens that cover natural soil (Tian et al. 2021). Infrastructure, whether green (natural) or gray (built), should be integrated to support urban functions while also providing habitats for wildlife, ensuring that development works in harmony with nature.

The second principle is to embed educational opportunities into places. This involves using design elements, signage, and wayfinding systems to create walkable environments that promote healthy behaviors, positive social interactions, and a deeper connection to nature. By making ecological processes visible and understandable—such as through interpretive trails or community gardens—residents can better appreciate the environmental impacts of their actions and the importance of natural systems. This principle not only fosters environmental stewardship but also strengthens community bonds and encourages sustainable lifestyles.

The third principle is to cultivate nature as a dynamic, co-created outcome. SEPL&S is not about preserving pristine wilderness but rather about fostering landscapes where human activities and natural processes interact in mutually beneficial ways. The sustainability of these landscapes in providing ecosystem services, such as clean water, pollination, and climate regulation, relies on healthy, intentional interactions with humans. This principle emphasizes the role of people as active participants in shaping and maintaining these landscapes, whether through sustainable agriculture, urban forestry, or habitat restoration, as exemplified in the Satoyama landscape history in Japan (Takeuchi et al. 2012). By viewing nature as a cultivated and collaborative outcome, we can create landscapes that are both productive and resilient, supporting both ecological health and human well-being.

Effective implementation of these placemaking principles in the Galápagos will require action-oriented research and policy experimentation tailored to the archipelago's unique context. An urgent priority is to assess the adequacy and quality of public spaces and amenities to promote inclusive and healthy lifestyles in human settlements, especially given that the region has one of the highest obesity rates in Ecuador. Identifying gaps in public infrastructure—such as parks, sidewalks, and recreational areas—to ensure they meet the diverse needs of all residents, regardless of age, gender, or physical ability, is both a critical and complex challenge. Mobilizing citizen science initiatives, involving residents in the planning process, and partnering with non-profit organizations and educational institutions could offer an efficient and participatory solution. A promising example is a collaborative project being developed by the municipality of Puerto Baquerizo Moreno on San Cristobal Island, the University of San Francisco Quito (USFQ), and the University of Oregon, which seeks to redesign public spaces to enhance walkability, social interaction, and ecological integration.

Another inspiring example is Hacienda Tranquila, a local organization on San Cristóbal Island that operates a sustainable farm providing training in agroecological practices. This initiative not only encourages the cultivation of native species but also delivers economic benefits to the community, demonstrating how physical spaces can be designed to harmonize social and economic objectives. Similar models, such as Hacienda Montemar, El Chato Reserve, and Rancho Primicias, also integrate agritourism into their operations. The next challenge is to develop strategies for scaling up these positive placemaking efforts, creating public spaces and landscapes that foster healthier lifestyles, ecological resilience, and economic sustainability. By doing so, these initiatives can help transform the archipelago's human settlements into vibrant, inclusive, and regenerative environments that embody the principles of bioregionalism and socio-ecological production landscapes and seascapes (SEPLS).

Evidence Driven Approach for Co-adaptation Knowledge Co-production

The process of creating socio-ecological production landscapes and placemaking visions in the Galápagos is inherently driven by the co-adaptive relationship between the region's complex human and natural systems. This relationship underscores the reality that there is no perfect science—nor will there ever be—to allow for infallible predictions or management (Berkes 2009). Instead, the knowledge and intelligence guiding these efforts must be continuously gathered and refined through cyclical feedback loops, supported by carefully monitored sustainability evaluations that reinforce the interconnectedness of these systems. Adaptive management theories, which emphasize iterative learning and flexibility, inform the regional planning and placemaking approaches discussed here. These theories advocate for a continuous process of monitoring, learning, and adjusting strategies based on observed outcomes, ensuring that actions remain responsive to the dynamic and evolving nature of socio-ecological systems.

The principles of action guided by adaptive management theories emphasize collaborative decision-making, the integration of scientific and local knowledge, and the fostering of learning and knowledge sharing (Heylings and Bravo 2007) Adopting an experimental mindset is crucial, as it encourages the implementation of pilot projects and small-scale interventions to explore innovative solutions before scaling them up. This approach allows for iterative learning and adaptation, ensuring that strategies remain responsive to dynamic socio-ecological conditions. A valuable example of this can be found in Japan's ancient satoyama landscapes, which have evolved over thousands of years through careful observation, trial-and-error, and the accumulated wisdom of local communities (Takeuchi et al. 2012). These landscapes are diverse in spatial scope and dynamic in temporal terms, reflecting a deep understanding of the interplay between human activities and natural systems. Today, the lessons from satoyama underscore the importance of combining science, local participation, and evidence-based experimentation to acquire the knowledge needed to establish resilient and sustainable mechanisms.

Over the past two decades, the Galápagos region has witnessed the development of collaborative governance structures under the framework of the 1998 Galápagos

Special Law. While these advancements are promising, there is still significant potential to enhance the capacity of communities and institutions to tackle emerging challenges, particularly those related to climate change. For example, implementing more robust monitoring systems to track ecological, social, and economic indicators would provide essential data for informed decision-making. Additionally, creating platforms for sharing lessons learned and best practices among stakeholders could cultivate a culture of learning and innovation within organizations and communities. Initiatives like those undertaken by Hacienda Tranquila, which focus on sustainable agriculture and community engagement, could serve as pilot projects, showcasing how local actions can align with broader adaptive management objectives.

Adaptive management processes must be inclusive and can significantly benefit from knowledge co-production. This requires actively engaging diverse stakeholders, including local communities, scientists, policymakers, and practitioners, in identifying problems, gathering relevant information, and participating in planning and decision-making processes. One effective way to support knowledge sharing and co-production is through community-engaged education programs. Universities worldwide are increasingly drawn to the Galápagos as a natural classroom, with over 20 institutions currently operating programs in the region. Encouraging these universities to integrate community-engaged service-learning projects into their curricula could serve as a powerful strategy, leveraging the process to benefit local communities. A recent example of this approach is the collaborative summer field school involving students from the University of Oregon(UO), the University of Hong Kong (HKU), and Universidad San Francisco de Quito (USFQ). Through immersive learning, the students developed actionable proposals aimed at improving local environments (Global Education Oregon n.d.). Such initiatives highlight the potential of education and collaboration to address real-world challenges while fostering mutual learning and community empowerment. By fostering partnerships between academic institutions, local communities, and other stakeholders, the Galápagos can continue to serve as a model for sustainable development and adaptive management in the face of global environmental challenges.

Conclusion

Reflecting on the history of governance in the Galápagos, this chapter underscores the persistent tension between conflict and collaboration, as well as the interplay between international influences and local priorities. Moving forward, the region must embrace a multidisciplinary and transdisciplinary approach to sustainability science, tailored to the unique context of the Galapagos. This includes deepening our understanding of global drivers of change, such as globalization and climate variability, while remaining grounded in the specific needs of local communities and the ecological integrity of the archipelago. Regional planning and placemaking emerge as critical tools for creating socio-ecological production landscapes, providing a physical foundation that integrates social, economic, and ecological processes. These intertwined systems are essential for enabling the Galapagos to move forward sustainably, balancing human development with the preservation of its unique natural heritage.

Meaningful change will unfold gradually, shaped by human activities and mindful of the ecological processes that sustain the landscape. Establishing effective mechanisms for harmonizing human and natural systems is not a short-term achievement but a long-term endeavor. It requires identifying socio-economic and ecological linkages and aligning policy interventions with the dynamics of both systems. Participatory research and knowledge co-production will be essential in this process, ensuring that diverse voices are heard and integrated. Ultimately, the goal is to expand the resilience of nature while channeling human agency toward sustainable coexistence. By learning from the past and envisioning collaborative futures, the Galápagos can chart a path toward harmony, redefining sustainability planning and governance for generations to come.

References

Adams WM (2004) Against extinction: the story of conservation. Earthscan

Alomía Herrera I, Paque R, Maertens M, Vanacker V (2022) History of land cover change on Santa Cruz Island, Galapagos. Land 11(7):7. https://doi.org/10.3390/land11071017

Amparo Alban M (2020) Galapagos at risk: and legal options for it's protection under international environmental law. Inter-American Institute on Justice and Sustainability

Berkes F (2009) Community conserved areas: policy issues in historic and contemporary context. Conserv Lett 2(1):20–25. https://doi.org/10.1111/j.1755-263X.2008.00040.x

Berkes F, Folke C, Colding J (1998) Linking social and ecological systems: management practices and social mechanisms for building resilience. Cambridge University Press

Bremner J, Perez J (2002) Demographic dynamics, gender, and resource use in the Galapagos Islands. InterCoast, Winter:20-21. https://www.crc.uri.edu/download/2002_41_CRC_GenderPopulationEnvironment.pdf

Burgiel SW, Muir AA (2010) Invasive species, climate change and ecosystem-based adaptation: addressing multiple drivers of global change global invasive species programme. Global Invasive Species Program. https://doi.org/10.13140/2.1.1460.8161

Cáceres R, Pittman J, Castrejón M, Deadman P (2022) The evolution of polycentric governance in the Galapagos small-scale fishing sector. Environ Manag 70(2):254–272. https://doi.org/10.1007/s00267-022-01666-z

Castrejón M, Defeo O, Reck G, Charles A (2014) Fishery science in galapagos: from a resource-focused to a social–ecological systems approach. In: Denkinger J, Vinueza L (eds) The Galapagos marine reserve: a dynamic social-ecological system. Springer International Publishing, pp 159–185. https://doi.org/10.1007/978-3-319-02769-2_8

CDF (n.d.) Our history. Charles Darwin Foundation. Retrieved February 20, 2025, from https://www.darwinfoundation.org/en/about/our-history/?utm_source=chatgpt.com

Connell J (2018) Islands: balancing development and sustainability? Environ Conserv 45(2):111–124. https://doi.org/10.1017/S0376892918000036

de Haan FJ, Quiroga Ferri D, Walsh SJ, Bettencourt LMA (2019) Scales and transformative change: transitions in the Galapagos. In: Kvan T, Karakiewicz J (eds) Urban Galapagos: transition to sustainability in complex adaptive systems. Springer International Publishing, pp 43–58. https://doi.org/10.1007/978-3-319-99534-2_4

Elmqvist T, Andersson E, Frantzeskaki N, McPhearson T, Olsson P, Gaffney O, Takeuchi K, Folke C (2019) Sustainability and resilience for transformation in the urban century. Nat Sustain 2(4):267–273. https://doi.org/10.1038/s41893-019-0250-1

Epler B (2007) Tourism, the economy, population growth, and conservation in Galapagos. Charles Darwin Foundation

FAO (n.d.) Climate-Smart Agriculture | Food and Agriculture Organization of the United Nations. Retrieved February 21, 2025, from https://www.fao.org/climate-smart-agriculture/en/

Folke C (2004) Traditional knowledge in social–ecological systems. Ecol Soc 9(3). https://doi.org/10.5751/ES-01237-090307

Folke C (2006) Resilience: the emergence of a perspective for social–ecological systems analyses. Glob Environ Chang 16(3):253–267. https://doi.org/10.1016/j.gloenvcha.2006.04.002

Galapagos Conservancy (n.d.-a) History of Galápagos. Galápagos Conservancy. Retrieved February 10, 2025, from https://www.galapagos.org/about_galapagos/history/

Galapagos Conservancy (n.d.-b) Sustainable tourism. Galápagos Conservancy. Retrieved February 21, 2025, from https://www.galapagos.org/travel/sustainable-tourism/

Geddes P (1898) The influence of geographical conditions on social development. Geogr J 12(6):580–586. https://doi.org/10.2307/1774281

Geddes P (1915) Cities in evolution: an introduction to the town planning movement and to the study of civics, Williams

Gehl J (2013) Cities for people. Island Press

Gehl J (2015, July 9) Life between buildings. Island Press. https://islandpress.org/books/life-between-buildings

Global Education Oregon (n.d.) Sustainable cities and landscapes in the Galapagos | Global Education Oregon. Retrieved February 21, 2025, from https://geo.uoregon.edu/programs/americas-ecuador/sustainable-cities-and-landscapes-galapagos

González JA, Montes C, Rodríguez J, Tapia W (2008) Rethinking the Galapagos Islands as a complex social-ecological system: implications for conservation and management. Ecol Soc 13(2)

Herremans IM, Ambrosie L (2013) Evolution of social institutions in the journey towards sustainability: the case of the Galápagos Islands

Heylings P, Bravo M (2007) Evaluating governance: a process for understanding how co-management is functioning, and why, in the Galapagos marine reserve. Ocean Coastal Manag 50(3):174–208. https://doi.org/10.1016/j.ocecoaman.2006.09.003

Hoyman MM, McCall JR (2013) Is there trouble in paradise? The perspectives of Galapagos community leaders on managing economic development and environmental conservation through ecotourism policies and the special law of 1998. J Ecotour 12(1):33–48. https://doi.org/10.1080/14724049.2012.749882

Johnson B, Ko Y (2022) The APRU sustainable cities and landscape hub: a platform for collaborative knowledge production and action. In: Yang Y, Taufen A (eds) The Routledge handbook of sustainable cities and landscapes in the Pacific rim. Routledge, Taylor Francis Group

Kvan T, Karakiewicz J (eds) (2019) Urban Galapagos: transition to sustainability in complex adaptive systems. Springer International Publishing. https://doi.org/10.1007/978-3-319-99534-2

LaCondamine C (2022) Past and present of conservation in the Galapagos Islands. ArcGIS StoryMaps. https://storymaps.arcgis.com/stories/ba7b70fe12914d0c9f02a9d61dfbe96f

Lynch K (2008) The image of the city. MIT Press

Mačić V, Albano PG, Almpanidou V, Claudet J, Corrales X, Essl F, Evagelopoulos A, Giovos I, Jimenez C, Kark S, Marković O, Mazaris AD, Ólafsdóttir GÁ, Panayotova M, Petović S, Rabitsch W, Ramdani M, Rilov G, Tricarico E et al (2018) Biological invasions in conservation planning: a global systematic review. Front Mar Sci 5. https://doi.org/10.3389/fmars.2018.00178

Maestro M, Pérez-Cayeiro ML, Reyes H, Chica-Ruiz JA (2024) Analysis of the management of protected areas in Galapagos: 60 years after its declaration as a National Park. Sustainability 16(15):15. https://doi.org/10.3390/su16156532

Marques AL, Alvim ATB, Schröder J (2022) Ecosystem services and urban planning: a review of the contribution of the concept to adaptation in urban areas. Sustainability 14(4):4. https://doi.org/10.3390/su14042391

McHarg IL (1995) Design with nature. Perfection Learning Corporation

Mostafavi M, Doherty G, Askins K (eds) (2016) Ecological urbanism. Lars Muller Publishers

Nel R, Mearns KF, Jordaan M, Goethals P (2021) Towards understanding the role of islandness in shaping socio-ecological systems on SIDS: the socio-ecological islandscape concept. Eco Inform 62:101264. https://doi.org/10.1016/j.ecoinf.2021.101264

Quiroga D (2009) Crafting nature: the Galapagos and the making and unmaking of a "natural laboratory.". J Polit Ecol 16(1):123–140. https://doi.org/10.2458/v16i1.21695

Quiroga D (2013) Changing views of the Galapagos. In: Walsh SJ, Mena CF (eds) Science and conservation in the Galapagos Islands: frameworks & perspectives. Springer, pp 23–48. https://doi.org/10.1007/978-1-4614-5794-7_2

Raman R, Leal Filho W, Martin H, Ray S, Das D, Nedungadi P (2024) Exploring sustainable development goal research trajectories in small Island developing states. Sustainability 16(17):7463. https://doi.org/10.3390/su16177463

Raufflet E, Berkes F, Folke C (2000) Review of linking social and ecological systems: management practices and social mechanisms for building resilience. Conserv Ecol 4(2)

Raymond CM, Cebrián-Piqueras MA, Andersson E, Andrade R, Schnell AA, Battioni Romanelli B, Filyushkina A, Goodson DJ, Horcea-Milcu A, Johnson DN, Keller R, Kuiper JJ, Lo V, López-Rodríguez MD, March H, Metzger M, Oteros-Rozas E, Salcido E, Sellberg M et al (2022) Inclusive conservation and the Post-2020 global biodiversity framework: tensions and prospects. One Earth 5(3):252–264. https://doi.org/10.1016/j.oneear.2022.02.008

Reale N (2022, October 14) The Galápagos Islands: economy over ecosystems. Harvard International Review. https://hir.harvard.edu/the-galapagos-islands-economy-over-ecosystems/

Santader T, Gonzalez J, Tapia W, Araujo E, Montes C (2009) Research trend in Galapagos and its implication for the archipelagos

Satoyama Initiative (2010) Paris Declaration on the "Satoyama Initiative." https://satoyama--initiative.org/wp-content/uploads/2014/02/Paris-Declaration-on-the-Satoyama-Initiative.pdf

Scheyvens H, Mader A, Lopez-Casero F, Takahashi Y (2019) Socio-ecological production landscapes and seascapes as regional/local circulating and ecological spheres. IGES Issue Brief

Smart Growth America (n.d.) What is smart growth? Smart Growth America. Retrieved February 21, 2025, from https://smartgrowthamerica.org/what-is-smart-growth/

Snell HM (1996) Conservation CETS personal. Noticias de Galápagos 56:13–16

Takeuchi K, Brown RD, Washitani I, Tsunekawa A, Yokohari M (2012) Satoyama: the traditional rural landscape of Japan. Springer Science & Business Media

Tengö M, Brondizio ES, Elmqvist T, Malmer P, Spierenburg M (2014) Connecting diverse knowledge systems for enhanced ecosystem governance: the multiple evidence base approach. Ambio 43(5):579–591. https://doi.org/10.1007/s13280-014-0501-3

Thompson K, Duinker PN, Sherren K, Hayden A, Terashima M (2024) The ecosystem services concept in urban planning: the criteria for practical fit. Plann Pract Res 39(5):813–838. https://doi.org/10.1080/02697459.2024.2359784

Tian A, Zünd D, Bettencourt LMA (2021) Estimating rooftop solar potential in urban environments: a generalized approach and assessment of the Galápagos Islands. Front Susta Cities 3:632109. https://doi.org/10.3389/frsc.2021.632109

Town and Country Planning Association (n.d.) Garden City Principles. Town and Country Planning Association. Retrieved February 21, 2025, from https://www.tcpa.org.uk/garden-city-principles/

Uchiyama C, Ohno T, Kubo Y (2023) Review of adaptive governance (AG) research in small islands: critical dimensions and future directions. J Mar Island Cult 12(3). https://doi.org/10.21463/jmic.2023.12.3.05

UNESCO World Heritage Centre (1998) UNESCO World Heritage Centre – State of Conservation (SOC 1998) Galápagos Islands (Ecuador). UNESCO World Heritage Centre. https://whc.unesco.org/en/soc/2191/

UNESCO World Heritage Centre (n.d.) Galápagos Islands. UNESCO World Heritage Centre. Retrieved February 20, 2025, from https://whc.unesco.org/en/list/1/

UN-Habitat (2015) International Guidelines on Urban and Territorial Planning (IG-UTP) Handbook | UN-Habitat. https://unhabitat.org/international-guidelines-on-urban-and-territorial-planning-ig-utp-handbook

UN-Habitat (2019) Urban-rural linkages: guiding principles | UN-Habitat. https://unhabitat.org/urban-rural-linkages-guiding-principles

United Nations (n.d.) THE 17 GOALS | Sustainable Development. Retrieved February 21, 2025, from https://sdgs.un.org/goals

Valdivia G, Wolford W, Lu F (2014) Border crossings: new geographies of protection and production in the Galápagos Islands. Ann Assoc Am Geogr 104(3):686–701. https://doi.org/10.1080/00045608.2014.892390

Waldheim C (ed) (2006) The landscape urbanism reader. Princeton Architectural Press

Waldheim C (2016) Landscape as urbanism: a general theory. Princeton University Press

West P, Igoe J, Brockington D (2006) Parks and peoples: the social impact of protected areas. Annu Rev Anthropol 35(1):251–277. https://doi.org/10.1146/annurev.anthro.35.081705.123308

Yang Y, Taufen A, Tochen R (2022) Learning spaces of policy mobility for sustainable cities and landscapes: the role of researchers and educators. In: Yang Y, Taufen A (eds) The Routledge handbook of sustainable cities and landscapes in the Pacific rim. Routledge, Taylor Francis Group, pp 22–35

Epilogue: Future Galapagos

Stephen J. Walsh and Carlos F. Mena

Introduction

Born of fire, formed through volcanism, isolated by geography, shaped by oceanic origins, famed for adaptation and evolution, and now influenced by the expanding human dimension, the Galapagos Islands of Ecuador are a testament to the natural and anthropogenic forces of change that inspire global visitors, attract migrants, and engage interdisciplinary scientists. These forces and factors of change shape and reshape the social, terrestrial, and marine subsystems as well as the human, cultural, and environmental contexts of past, present, and future conditions. Change across space, time, and contexts has been a hallmark of the Galapagos Islands from their earliest beginnings. Plate tectonics and continental drift are global forces of change that have formed these islands, given them their shape and structure, and created niche spaces for colonization by plants whose seeds arrive carried by the wind, pushed by marine currents, and transported by animals that are good floaters, flyers, or swimmers.

The juxtaposition of the Galapagos Islands on the volcanic platform that they reside, climatic conditions that influence patterns and processes of life, interplay of cold and warm ocean currents that attest to high levels of marine productivity and species diversity, and the fascinating symbiotic relationships among species have

S. J. Walsh
Department of Geography & Environment, University of North Carolina at Chapel Hill, Chapel Hill, NC, USA
e-mail: swalsh@email.unc.edu

C. F. Mena
Galapagos Science Center, San Cristobal Island, Universidad San Francisco de Quito, Quito, Ecuador
e-mail: cmena@usfq.edu.ec

J. Dunn Insua et al. (eds.), *Human Galapagos*, Social and Ecological Interactions in the Galapagos Islands, https://doi.org/10.1007/978-3-032-08800-0

influenced the evolving nature of the Galapagos Islands as well as the native and endemic species that reside there. With the illusion of fresh water, many areas of the Galapagos have been rendered uninhabitable for humans, but ripe for species colonization of inaccessible habitats. Plankton-rich ocean waters provide a wealth of favorable conditions for the introduction of warm- and cold-water species. A world devoid of large mammals, key herbivores in the Galapagos are reptiles.

With the arrival of human visitors to the Galapagos beginning in the 1500s, the human dimension continues to expand, accelerated when the Galapagos Islands became part of Ecuador in 1832. Today, approximately 40,000 residents live in the Galapagos, mostly supported by the burgeoning tourism industry that draws over 300,000 visitors annually to the "Enchanted Islands." The Galapagos Islands are also a scientific laboratory for the study of human-environment interactions, made famous by Charles Darwin's passage through the archipelago in 1835 on-board the *HMS Beagle*. Today, the uninterrupted exploration of the Islands by a diverse, interdisciplinary team of scientists from institutions located far and wide, includes the Galapagos Science Center on San Cristobal Island founded in 2011 as a program of education, research, and community engagement, jointly developed and operated by the Universidad San Francisco de Quito, Ecuador and the University of North Carolina at Chapel Hill, USA.

Exhibiting both vulnerability as well as resilience, the Galapagos Islands offer a relatively harsh natural environment to welcome its early and contemporary residents and visitors alike. For instance, while tourism helps the local economy, with nearly 80% of residents working in the tourism industry, it also potentially harms the local resources and the unique species that visitors seek to enjoy. Today, more non-native species exist in the Galapagos Islands than native ones and the native and endemic species suffer from the increased use-pressures that are attendant with global tourism. At the same moment that tourists and residents flock to the Galapagos Islands for access to nature and jobs in tourism, challenges to sustainable agriculture, urbanization, fisheries, and economic development have become important stressors on current conditions and future possibilities. Particularly troubling is when social, terrestrial, and marine subsystems in the Galapagos are perturbed by recurring forces of change, such as, ENSO events (El Nino Southern Oscillation) or a pronounced increase in tourism and economic development, coincident with the needs of visitors and residents.

The fragile balance between protection and development of the Galapagos Islands suggests a future where species as well as human adaptation to change will be paramount and where alternate realities must be considered for this special place to retain its magical character for sustainable life for the original residents, plants and animals, as well as for the relatively recent human collective. The "Galapagos Paradox" is often cited to describe the conflicts between development and conservation, as people migrate to the Galapagos for jobs in the tourism industry and at the same time tourists arrive in the islands to interact with nature, particularly, with the iconic and emblematic species that are the ecological "signatures" of the Galapagos Islands. Designated as a World Heritage Site, a National Park, and a Marine Reserve, the uniqueness of the environments across the archipelago adds to the reputation of the islands, encouraging touristic visitors and working residents to permanently or temporarily migrate to the islands, thereby, threatening the islands' very existence and their future sustainability.

Change in the Galapagos

Several papers have been published that examine plausible scenarios of change for the Galapagos Islands. A significant number of these papers have relied upon agent-based models and dynamic systems models to study coupled human-natural systems and possible perturbations to them because of population-environment interactions and specific exogenous and endogenous forces of change. Walsh and Mena (2016) generated models of the Galapagos Islands, rooted in complexity theory, by describing elements, relationships, and processes that enhance our understanding of human-environment interactions. The general approach formalized knowledge of how systems in the Galapagos operate and the manner that key elements are linked in coupled human-natural systems through formal rules, relationships, and rates of exchange between social, ecological, and biophysical features, derived through statistical functions and/or functions specified in theory or practice. The models are data rich, calibrated and validated, and interpreted within empirical and theoretical contexts. The derived models consider ecosystems as complex adaptive systems because their macroscopic properties emerge from the interactions among the individual components of the ecosystem. Not only are biological components of ecosystems subject to selection, but their macroscopic properties may also feed back into the system and affect its future development (Levin 1998).

Dynamic systems models manage complex feedback systems through the collection of interacting elements that function together in a specified system. Feedback loops are the key to understanding system dynamics. Properties in the models include quantities that vary over time, variability is described causally, and influences are contained within a closed-system feedback loop. The dynamic systems models are used to study economic development in the Galapagos Islands associated with the tourism industry and the feedback mechanisms that impact environmental conditions with the goal of ecosystem sustainability. Conversely, an agent-based model is a computational approach for experimenting with complex systems that involves sets of heterogeneous agents and frameworks for simulating each agent's set of decisions and interactions within a dynamic environment. Such systems evolve through time, using a description of the changing behaviors of individual agents and the environment to link people and the environment and to examine alternate decision-making, learning, adaptation, and change (Malanson and Walsh 2015). The generation of the dynamic systems models by Walsh and Mena (2016) are designed to examine future scenarios of human-environment interactions with a particular focus on the forces of change in tourism and residential growth, environmental degradation, household demographics, and community infrastructure variables. The purpose of the agent-based models are to simulate the decision-making processes of fishers in the Galapagos Islands with regards to employment choices and an alternate household livelihood strategy that moves those qualified into tourism (full- or part-time), given a variety of circumstances, desires, and personal characteristics.

Among the findings reported by Walsh and Mena (2016), the primary demographic variable driving the rapid residential demographic growth in the Galapagos Islands is population migration, with most migrants coming from the Ecuadorian mainland. Future growth in the resident population is tied to growth in tourism. If tourism continues to grow at the pace of the past two decades in several years, there will be no open space to ensure that every tourist visiting the Galapagos experiences nature tourism at a high standard of quality. In parallel, social, administrative, and organizational systems in the Galapagos have not demonstrated the ability to evolve at the same rate that has seen the expansion of tourism, creating imbalances and failures as well as interactions that are negatively associated with the tourism sector and amplified for fragile and sensitive island ecosystems occurring within the Galapagos (Mena et al. 2020a). Further, the models reported by Walsh and Mena (2016) suggest a Galapagos that includes more urbanization, expansion of land-based tourism, and environmental change, primarily expansion of invasive species and reduction in sustainable agriculture. In addition, using stocks, flows, and rates of change, national and international tourism and land- and boat-based tourism affect the evolution of physical capital, social capital, and natural capital that are integrated and related to changes in ecosystem goods and services (Hernandez and Leon 2007).

An outcome from both models is the central role of tourism in the development of the Galapagos, i.e., the driving force in determining the dynamics of change in all other employment sectors (fisheries, agriculture, and services) of the islands, particularly, generating high flows of migration from the mainland to the islands, increasing the introduction of invasive species, accelerating the consumption of resources, increasing land cover/land use change, and placing considerable pressure on basic services, e.g., power generation, freshwater availability, waste collection and recycling, education, healthcare, and governance (Pasmino et al. 2018). For the fisheries sector, the agent-based models effectively simulate the decision-making processes of fishers relative to the pushes out of fishing and pulls into tourism, given expenses, income, wealth, and preparation for alternate employment possibilities. The dynamic systems models emphasized the processes related to population growth, including natural residential growth and in-migration that drives changes in different components of the Galapagos social and environmental systems.

With a substantial emphasis on tourism management in the Galapagos Islands as a recurring theme reported by diverse researchers, Pizzitutti et al. (2014) used agent-based models to supplement the literature about the Galapagos Islands by enriching our understanding of system behaviors and the adaptive behavior of individual agents to represent the touristic market conditions in the Galapagos Islands, focusing on touristic offers, reservations, and touristic activities. The model was based on an individual-based representation of tourists' consumptive preferences and touristic accommodation offers in the Galapagos Islands. Further, the model was used to generate “what if” scenarios of change to study how emergent patterns in the touristic market in the Galapagos are affected by changes in the environment, illustrating how the model can be used to enable public policy makers to explore the consequences of their decisions. In addition, a decision-support system based on a dynamic systems model was used to examine tourism management in the Galapagos Islands. Each subsystem in the models was examined through hypothesized scenarios of tourism growth, associated with different residential population expansions levels.

In related work, Pizzitutti et al. (2017) used a participatory approach embedded within dynamic systems models to explore scenarios of tourism management within the islands. They relied upon a decision-support system of multiple stakeholders in the Galapagos to assess system behaviors and the generation of graphical representations of the impacts of tourism growth on associated residential population growth. Model outcomes point towards the growing number of introduced species that threaten island ecosystems in the Galapagos, and the rapid saturation of the tourism reception capacity. Touristic enrichment in the Galapagos is limited by the limited number of high interest terrestrial and marine visitation sites for use by tourists, whose intention is to interact with a spectacular natural environment that the Galapagos Islands have become famous and world-renowned.

Lastly, Pizzitutti et al. (2020) developed the dominant trends in land cover/land use change in the Galapagos, mapped by satellite systems and derived multispectral imagery. The mapped patterns were linked to the direct and indirect socio-economic drivers of change, characterized by global tourism, demography, population migration, economic development, and labor market conditions in the Galapagos Islands. The development of the models were motivated by the critical importance of human modifications of the environment and the impacts of worldwide biodiversity and ecosystem health on global climate change, and the corresponding changes to the living conditions of many human societies (Steffen et al. 2004). Furthermore, by directly hampering ecological sustainability and ecosystem goods and services, human impacts on the environment create a suite of multi-scale risks to the natural environment that may be incapable of sustaining societal needs now and into the future (Diaz et al. 2019). In the Galapagos, using Santa Cruz Island as the point of reference, Pizzitutti et al. (2020) found that under different tourism growth scenarios, the degraded areas of the island continued to expand to the point of doubling in size under exponential tourism growth. Further, most of the land degradation areas on the island are concentrated in the humid vegetation zone of the rural highlands, with urban expansion occurring primarily in the coastal zone. Finally, the model indicates that tourism growth in the Galapagos and the Ecuadorian national economy, the main exogenous drivers of social change, influenced demographic dynamics through pronounced impacts on immigration and emigration of people to and from the Galapagos Islands in response to economic conditions and employment opportunities in the islands and on the Ecuadorian mainland.

Possible Galapagos Futures

It has been well document for the Galapagos Islands and for several island settings around the globe that tourism is a dominant driver of environmental change, including both direct and indirect effects that severly impact island ecosystems and generate several risks and threats to island sustainability (Mena et al. 2020b). While there are many approaches reported in the literature to assess the social, terrestrial, and marine subsystems in the Galapagos and their integration, dynamic systems models and agent-based models are well designed to examine coupled human-natural systems, decision-making individuals, alternate household livelihood strategies, and learning and adaptation, tuned by internal and external forces of change and the role of feedback mechanisms on system behaviors. With the capability to generate scenarios of change over space and time, the models support policy development and assessment by creating "what if" scenarios of change for prescribed periods, places, and scales, often reported within a graphical representation in which variables, interactions, and behaviors are calibrated, validated, and generated within an experimental framework. Therefore, within the context of generating plausible and even likely outcomes of coupled human-natural systems, the following observations are offered for the Galapagos Islands and their possible future outcomes.

Protected for centuries through geographic isolation, the Galapagos Islands are now linked to the world through globalization in which people, goods, and services pass to and from the Galapagos Islands affecting residents and visitors. Through, social and environmental linkages and/or electronic and physical connections, the Galapagos Islands are tied to global places like never before. Whether it be through politics, economics, migration, or health issues, the islands respond to endogenous and exogenous forces of change in complex ways. Likely, the connection of the Galapagos Islands to near and distant places will only increase through time, bringing new opportunities for economic development and a suite of environmental challenges to this remote and fascinating place.

In 1997–1998, for instance, the world experienced an intense ENSO (El Nino Southern Oscillation) event that, in the Galapagos, challenged the social, terrestrial, and marine subsystems in fundamental ways. With a dramatic disruption of local upwelling, marine productivity was substantially reduced, thereby, severely damaging local fisheries and the corresponding employment in that historically important industry. Also, with increases in the associated rainfall levels, invasive plant species flourished and expanded their reach across the landscape, threatening local and endemic species. Several iconic species were impacted by a reduction in food resources, severely jeopardizing local populations, for instance, marine iguanas, flamingoes, Galapagos penguins, and Galapagos sealions. For species whose populations were already relatively small, the El Nino event pushed them to the very brink of existence, although some terrestrial species, such as, the giant tortoise fared well with more abundant vegetation. It has been forecast that the Galapagos Islands, operating at the epicenter of El Nino events, will realize a greater frequency and intensity of ENSO events in the future, thereby, placing several iconic species at risk within "Darwin's Paradise."

In April of 2007, also, for instance, Ecuadorian President Rafael Correa signed a decree that declared the Galapagos Islands "at risk" and "a national priority" for preservation. In June of that same year, UNESCO listed the Galapagos as an "endangered" World Heritage Site. These declarations offer some context to the national and international concern for the future of the Galapagos Islands, focusing on the increasing trends in tourism and the corresponding expansion of the resident population who migrate to the Galapagos Islands for jobs in the growing tourism industry. Other concerns included the expansion of invasive species, imports of food, products, and energy from the Ecuadorian mainland, and ineffective food security programs. In addition, the local population voiced longstanding concerns for improved healthcare, education, and water quality and availability. With household livelihood transitions of farmers (and fishers) to the tourism industry, land abandonment and reduced use occurred in the moist, agricultural highlands, chiefly caused by the lack of affordable and sustainable labor, spread of invasive species, and economic competition of farm products grown in the Galapagos with products grown on the Ecuadorian mainland and shipped to the islands. With local cultivation of farm products an expressed goal in the Galapagos to support food security issues, future growth in local agriculture is likely, but the question is whether the expanded cultivation of farm products in the islands can keep pace with local demand of an increasing tourism and residential population as well as the economic competitiveness of local versus imported agricultural products into the future.

Due to the tourism industry and the associated growth of the resident population, the status of urban expansion is alarming. It is not only the increased densification of structures, but it is also the considerable stress that is being placed on the urban infrastructure, particularly, water and sanitation, energy consumption, healthcare delivery, and education. With an expansion of land-based tourism and island hopping, the continued expansion of urban centers will elevate future concerns for the protection of fragile ecosystems and the sustainability of islands throughout the archipelago. Boat-based tourism is also likely to expand with the demand for larger and more luxurious vessels, coincident with the enhancement of port facilities to support touristic venues. With approximately 150 tourist visitation sites distributed throughout the archipelago, more-or-less evenly divided between terrestrial and marine, the enhancement of current sites and the development of additional sites is necessary to ensure a high-quality nature experiences of global visitors and residents. It is likely that additional land sites will be cultivated in the moist highlands in association with farmers seeking to diversify into tourism related experiences, such as, the marketing of giant tortoises on farm properties and the establishment of bird habitats for touristic viewing. Considering farms as units of protection and conservation will be vital in this endeavor to more fully integrate farms and farmers into the tourism industry. Sustainable agriculture and food security will take a more diversified form in which farmers subsidize their household incomes through tourism by transitioning working farms into demonstration and visitation sites, and by concentrating mechanized cultivation on farms that are geographically joined for a more robust engagement of farm technologies limited by the relatively small size and disjointed organization of current farms. Continued development of artificial reservoirs in the highlands will help ensure sustainable agriculture for local consumption and the attraction and retention of iconic species.

In recent years, the character of tourists visiting the Galapagos Islands has been changing. Historically, visitors to the Galapagos arrived with the intention to conserve and protect fragile ecosystems and to observe or participate in actual programs that ensure ecosystem sustainability. Today, an increasing number of visitors are intent on visiting an exotic destination to engage in recreational pursuits, such as, surfing, diving, and sport fishing. This change in the character of tourists has led to the commercialization of activities that are linked to recreation, thereby, accelerating this transition to alternate experiences.

Lastly, the Galapagos Islands remain a strong attractor for research and education institutions, particularly, national and international universities, foundations, and institutes as well as the engagement of faculty and student scientists from individual campuses. The USFQ-UNC Galapagos Science Center, located on San Cristobal Island, is dedicated to research, education, and community engagement. With a host of associated interdisciplinary scientists affiliated with the Galapagos Science Center (GSC), hundreds of projects have been successfully conducted in the Galapagos utilizing its professional staff, 20,000 square feet of building space, and laboratories that focus on marine ecology, terrestrial ecology and microbiology, genetics and molecular science, and spatial analysis and modeling. The GSC will continue to be a force within the Galapagos, engaging local people through the formation of a citizen science corps and related community initiatives as well as attracting diverse scholars through research collaborations. In addition, USFQ is developing a Santa Cruz campus designed to support residents through training in specialized skills and knowledges that are directly applicable to trades and services in tourism as well as integrated studies in ecosystem goods and services for a healthy and sustainable environment. Within this context, the GSC has launched several projects that are framed within a *One Health* approach in which diverse, interdisciplinary, and collaborative scholars seek to attain optimal health for people, plants, animals, ecosystems, and the environment more broadly through integrated studies and actions that sustain island ecosystems now and into the future. The central goal is for the *One Health* approach to increasingly draw scientists together for the improved understanding of integrated systems as perturbed by external and internal shocks and forces of change to guide coupled human-natural systems towards future sustainability.

Conclusions

There are many questions and concerns about the future of the Galapagos Islands and how the social, terrestrial, and marine subsystems will adapt to change as new pathways into the future are charted. The central question is whether the "*Galapagos*

Islands can survive?" and what will the Galapagos look like, feel like, and represent as the islands iteratively respond to exogenous shocks and endogenous dynamics that create new challenges and new opportunities, embraced or rejected through alternate scenarios of the future. An associated question is "*How will social systems react to change?*" particularly, organizations, including the Galapagos National Park. So too, "*Might the coupled human-natural systems in the Galapagos attain a heighten resilience to change and more effectively shape its future in a sustainable manner?*"

Creative public policy is necessary to chart a sustainable future for the Galapagos. Financial investments, effective governance, integrated and comprehensive assessments of current conditions and future trajectories are vital to the future of the Galapagos. Participatory management has been successfully implemented to address an array of troublesome issues in the Galapagos, but other approaches such as dynamic systems modeling and agent-based models need further development, testing, and evaluation, so they might be conducted within a "what if" policy context. While the fragility and attraction of the Galapagos Islands have generally motivated their assessment through visions and activities that look inward and towards the Galapagos, placing the Galapagos Islands within a global context for comparison and guidance from the shared experiences and outcomes discerned for similarly threatened island settings is an approach that merits significant attention.

In short, the protection and conservation of the Galapagos Islands demands wise stewardship of its resources now and into the future. The critical challenges of tourism and population growth are fundamental to a sustainable future. How change is shaped, guided, informed, and interpreted will determine the kind of future that the Galapagos will attain. As Charles Darwin stated, "*It is not the strongest species that survive, nor the most intelligent, but the most adaptable to change.*"

References

Diaz, J., Settele, E.S., Brondizio, E.S., Ngo, H.T., Gueze, M., Agard, J. & Zayas, C.N. (2019). Report of the planetary of the intergovernmental science policy platform on biodiversity and ecosystem services on the work of the seventh session, Bonn, Germany

Hernandez JM, Leon CJ (2007) The interactions between natural and physical capitals in the tourist lifecycle model. Ecol Econ 63:184–193

Levin S (1998) Ecosystems and the biosphere as complex adaptive systems. Ecosystems 1(5):431–436

Malanson GP, Walsh SJ (2015) Agent-based models: individuals interacting in space. Appl Geogr 56:95–98

Mena CF, Paltan HA, Benitez FL, Sampedro C, Valverde M (2020a) Threats of climate change in small oceanic islands: the care of climate and agriculture in the Galapagos Islands, Ecuador. In: Walsh SJ, Riveros-Iregui D, Arce-Nazario J, Page PH (eds) Land cover and land use change on islands – social and ecological threats to sustainability. Springer-Nature, Cham. https://doi.org/10.1007/978-3-030-43973-6_5

Mena CF, Quiroga D, Walsh SJ (2020b) Threats to sustainability in the Galapagos Islands: a social-ecological perspective. In: Sarmiento FO, Frolich LM (eds) International handbook of geography and sustainability. Edward Elgar Publishing, Cheltenham

Pasmino A, Serrao-Neumann S, Choy D (2018) Towards comprehensive policy integration for the sustainability of small islands: a landscape-scale planning approach for the Galapagos Islands. Sustainability 10:1228. https://doi.org/10.3390/su10041228

Pizzitutti F, Mena CF, Walsh SJ (2014) Modelling tourism in the Galapagos Islands: an agent-based modeling approach. J Artif Soc Soc Simul 17(1):1–14

Pizzitutti F, Walsh SJ, Rindfuss RR, Gunter R, Quiroga D, Tippett R, Mena CF (2017) Scenario planning for tourism management: a participatory and system dynamics model applied to the Galapagos Islands of Ecuador. J Sustain Tour 25(8):1117–1137

Pizzitutti F, Brewington L, Walsh SJ (2020) Human and natural environments, Island of Santa Cruz, Galapagos: a model-based approach to link land cover/land use changes to direct and indirect socio-economic drivers of change. In: Walsh SJ, Riveros-Iregui D, Arce-Nazario J, Page PH (eds) Land cover and land use change on islands – social and ecological threats to sustainability. Springer-Nature, Cham. https://doi.org/10.1007/978-3-030-43973-6_8

Steffen W, Sanderson RA, Tyson PD, Jager J, Matson PA, Moore B, Wasson RJ (2004) Global change and the earth system. Springer, Berlin

Walsh SJ, Mena CF (2016) Interactions of social, terrestrial, and marine sub-systems in the Galapagos Islands, Ecuador. Proc Natl Acad Sci 113(51):14536–14543

J. Dunn Insua et al. (eds.), *Human Galapagos*, Social and Ecological Interactions in the Galapagos Islands, https://doi.org/10.1007/978-3-032-08800-0

MIX
Papier aus verantwortungsvollen Quellen
Paper from responsible sources
FSC® C105338

If you have any concerns about our products,
you can contact us on
ProductSafety@springernature.com

In case Publisher is established outside the EU,
the EU authorized representative is:
Springer Nature Customer Service Center GmbH
Europaplatz 3, 69115 Heidelberg, Germany

Printed by Libri Plureos GmbH
in Hamburg, Germany